Coachbuilt Cars

Jaguar XK 120 Supersonic

Also published by Porter Press International

The Jaguar Portfolio
Ultimate E-type – The Competition Cars
Jaguar E-type – The Definitive History (2nd edition)
Original Jaguar XK (3rd edition)
Jaguar Design – A Story of Style
Saving Jaguar

Great Cars series
No. 1 – Jaguar Lightweight E-type – The autobiography of 4 WPD
No. 2 – Porsche 917 – The autobiography of 917-023
No. 3 – Jaguar D-type – The autobiography of XKD 504
No. 4 – Ferrari 250GT SWB – The autobiography of 2119 GT
No. 5 – Maserati 250F – The autobiography of 2528
No. 6 – ERA – The autobiography of R4D
No. 7 – Ferrari 250GTO – The autobiography of 4153 GT
No. 8 – Jaguar Lightweight E-type – The autobiography of 49 FXN
No. 9 – Jaguar C-type – The autobiography of XKC 051
No. 10 – Lotus 18 – The autobiography of Stirling Moss's '912'
No. 11 – Ford GT40 – The autobiography of 1075
No. 12 – Alfa Romeo Monza – The autobiography of the celebrated 2211130
No. 13 – Bugatti Type 50 – The autobiography of Bugatti's first Le Mans car

Exceptional Cars series
No. 1 – Iso Bizzarrini – The remarkable history of A3/C 0222
No. 2 – Jaguar XK120 – The remarkable history of JWK 651
No. 3 – Ford GT40 MkII – The remarkable history of 1016
No. 4 – The First Three Shelby Cobras
No.5 – Aston Martin Ulster – The remarkable history of CMC 614
No. 6 – Maserati 4CLT – The remarkable history of chassis no. 1600
No. 7 – Ferrari 250 LM – The remarkable history of 6303

Porter Profiles
No. 1 – Austin Healey – The story of DD 300

Original Scrapbooks
Barry Cryer Comedy Scrapbook
Graham Hill Scrapbook 1929–1966
Martin Brundle Scrapbook
Mini Scrapbook
Murray Walker Scrapbook
Stirling Moss Scrapbook 1929–1954
Stirling Moss Scrapbook 1955
Stirling Moss Scrapbook 1956–1960
Stirling Moss Scrapbook 1961

Deluxe leather-bound, signed, limited editions with slipcases are available for most titles.
Books available from retailers or signed copies direct from the publisher.

To order simply phone +44 (0)1584 781588, fax +44 (0)1584 781630, visit the website or email sales@porterpress.co.uk

Keep up to date with news about current books and new releases at:
www.porterpress.co.uk

Coachbuilt Cars

Jaguar XK 120 Supersonic

Richard Heseltine

Porter Press International

© **Porter Press International**

All rights reserved. No part of this publication may be reproduced,
stored in a retrieval system or transmitted, in any form or by any means,
electronic, mechanical, photocopying, recording or otherwise,
without prior permission in writing from the publisher

First published in September 2019

978-1-907085-82-6

Published by
Porter Press International Ltd
Hilltop Farm, Knighton-on Teme, Tenbury Wells, WR15 8LY, UK
Tel: +44 (0)1584 781588 Fax: +44 (0)1584 781630

sales@porterpress.co.uk
www.porterpress.co.uk

Edited by Giles Chapman
Design & Layout by Andrew Garman
Printed by Gomer Press Ltd

COPYRIGHT
We have made every effort to trace and acknowledge copyright holders and we apologise
in advance for any unintentional omission. We would be pleased to insert the appropriate
acknowledgement in any subsequent edition.

Contents

Introduction — 7

Chapter 1 — 8
The car: a considered overview

Chapter 2 — 24
The coachbuilder: the birth, decline and rebirth of Ghia

Chapter 3 — 38
The designer: the complex genius of Giovanni Savonuzzi

Chapter 4 — 52
Under the bonnet: the work of tuning ace Virgilio Conrero

Chapter 5 — 58
Firebrand: Luigi 'Gigi' Segre and his American adventures

Chapter 6 — 62
At the sharp end: Felice Mario Boano, coachbuilding workhorse

Chapter 7 — 66
Supersonic: origins of the species

Chapter 8 — 78
Jaguar XK 120: foundations of a legend

Chapter 9 — 84
Latin interpretations: the other Jaguar XKs with Italian bodywork

Chapter 10 — 90
Jaguar XK 120 Ghia Supersonic: the life and times of chassis 679768

Acknowledgements — 106

Index — 107

Introduction

Like most journeys into the unknown, this one began with an idea.

Giovanni Savonuzzi's shyness masked a brilliant mind and a ferociously independent streak. He revelled in creating a new kind of competition tool: a GT car aimed squarely at securing honours in the 1953 Mille Miglia road race across Italy. This brave new pinnacle of engineering would be lightweight and aerodynamically efficient, pushing the boundaries of what was possible and what was permissible.

Built in partnership with tuning legend Virgilio Conrero, the end product was breathtaking to behold. Notionally a reworked Alfa Romeo 1900, it emerged as a study in streamlining that employed components sourced from here, there and everywhere. It represented an alluring alchemy of grace and otherworldliness that screamed Jet Age. Unfortunately, the Mille Miglia bid ended in failure and disaster, this one-off 'special' being reduced to a heap of molten metal after it was ravaged by fire during the event.

That wasn't the end of the tale, however. The tragedy marked the beginning of one of the more convoluted tales in coachbuilding lore. The Supersonic's Savonuzzi-penned outline was too daring – too beautiful – to become a mere footnote in it. What could so easily have been just another striking but ill-starred 'etceterina', the sort of car forgotten by all but those who specifically seek out the obscure and esoteric, took on an intriguing new role. It became shorthand for all that was starry-eyed about car design. That, and the mastery of metal-manipulating Italian artisans, thanks to a limited run of replicas crafted by Italy's Carrozzeria Ghia.

The Supersonic series of cars became globally famous in the mid-1950s, inspiring automotive stylists in Europe, North American and beyond. The construction of these boutique GTs also brought in much-needed income for Ghia. Only a few years earlier, the Turin firm had been reduced to making pots and pans, window blinds and bicycle frames; anything to make ends meet as Italy dug itself out of the rubble at the end of the Second World War. The Supersonic in any of its several guises marked the emergence of the coachbuilder as a global influencer, with more than one giant Detroit carmaker beating a path to Ghia's door. And that's even before one considers motor manufacturers based closer to home. Accordingly, the Supersonic, in any of its incarnations, is important. These days, survivors are feted at *concours d'élégance* events, much as they were more than half a century ago.

When we first mooted this *Coachbuilt Cars* series of books, there was much discussion within Porter Press over which car should act as a template. It needed to be an exemplar of the coachbuilder's art; something that broke moulds and pushed boundaries. There was only one candidate on which we could all agree. It simply had to be a Supersonic. But not just any Supersonic.

Only 19 cars were produced by Ghia, based variously on Fiat 8V, Jaguar XK 120 and Aston Martin DB2/4 running gear. The Jaguar Supersonic was the most powerful of them all, thanks to its high state of engine tune. This remarkable machine has adorned covers of publications the world over and been garlanded at prestigious motoring events spanning several decades.

What follows is a full account of the car's history, from its birth in 1954 to the present day. It is the first time the extraordinary story has been fully told.

In order to make sense of the tangled circumstances that made its existence possible, the story makes its way through the 'building blocks' underpinning the Jaguar Supersonic, examining the companies, the key people, the donor car and the contemporary alternative designs. Threads of Italian, British and North American car culture are gradually drawn together in an automotive design story unlike any other. Once explained and interwoven, the history of the individual car that has made this book possible comes as the glorious climax to a particularly complex narrative.

I hope this book will, therefore, lead to an even greater appreciation of this Anglo-Italian icon and the men who made it happen.

Richard Heseltine
Shropshire, England
January 2019

Chapter One
The car: a considered overview

How do you improve on perfection? The Jaguar XK 120 caused a furore when introduced at the 1948 London International Motor Show at the capital's Earl's Court exhibition centre. It was achingly beautiful, and fast too, the '120' part of the nomenclature denoting its claimed top speed in miles-per-hour. It entered into legend in an instant and helped establish the Coventry marque on the world stage, before spawning legends such as the Le Mans 24 Hours-conquering C- and D-type sports-racing cars.

Most customers were happy for their Jaguars to be Jaguar-shaped, but not the mysterious Lyonnais businessman who ordered this car with coachwork by Carrozzeria Ghia. One of only three XK 120s bodied by the ambitious Turin design house, chassis number 679768 appeared at several major motor shows during the 1950s, in addition to keenly contested *concours* events.

The narrative is far from linear, however. The first owner failed to pay his bills before seemingly vanishing, leaving a trail of angry creditors in his wake. The car was then placed in storage, where it remained untouched before finally being disinterred at the end of the 1960s.

This fabulous machine in time was acquired by the co-founder of France's Jaguar Drivers' Club and went on to feature prominently in two private museums dedicated exclusively to the marque. A subsequent keeper initiated a sympathetic, no-expense-spared restoration, the end result wowing the judges at the prestigious Louis Vuitton Concours at Parc de Bagatelle, Paris in September 1996, where the car claimed the top award. A year later, custodianship passed to a different owner who also showed the car to appreciative audiences before the current keeper, William E. 'Bill' Heinecke, acquired it in 2015. The American-born Thai enthusiast (and former Macau Grand Prix entrant/racer) has merely continued from where his predecessors left off, sharing his enviable new possession with the public at a raft of prestigious events.

The car: a considered overview

● Kinetic sculpture: Giovanni Savonuzzi's sublime Supersonic outline raised the bar for jet-set styling in the 1950s. Several copyists followed in its wake but they were mostly pale imitations

The car: a considered overview

● It's hard to believe the Supersonic line was originally conceived with motor sport in mind. This fabulous Jaguar XK 120 variant married *à la mode* styling to high-performance.

The car: a considered overview

Jaguar XK 120 Supersonic

The car: a considered overview

● Far left: spare Borrani wire wheel and tool kit eat into the available luggage room, so the car is a little hampered for grand touring. Left: the stylish cabin is almost restrained by comparison with the *outré* exterior; the gearlever is notably slender.

The car: a considered overview

● One of Enrico Nardi's timelessly elegant wood-rim steering wheels fronts the Supersonic's duotone dashboard; the speedometer reads to 240km/h (150mph).

Jaguar XK 120 Supersonic

The car: a considered overview

● Jewel-like Virgilio Conrero-tuned Jaguar straight-six with its triple Weber carb set-up was the only example of the enduring XK engine ever modified by 'The Magician'.

Jaguar XK 120 Supersonic

The car: a considered overview

● The decorative emblems of the car are a metallic wonderland of detail, which even includes Ghia's own interpretation of the Jaguar badge among the complicated insignia; the panelbeating artistry around the rear wings, right, is breathtaking.

Jaguar XK 120 Supersonic

Chapter Two
The coachbuilder: the birth, decline and rebirth of Ghia

Scroll back half a century and Turin was to coachbuilding what Florence was to the Renaissance: crucible, arbiter and patron.

Dozens of *carrozzerie* (coachbuilders) operated out of workshops large and small, most with names bulging with vowels that sound impossibly exotic to non-native-speakers. Many of these now-legendary coachbuilders were clustered together, most within shouting distance of Fiat's enormous Lingotto plant, but their fame spread far beyond the city's borders. Stylists were rarely choked by caution, and hammer-wielding artisans translated their visions into full-scale reality with consummate skill. The speed with which they worked was incredible, so wealthy customers didn't have to wait long to bask in the reflective glow of all this creative brilliance.

Few coachbuilders were as daring and dynamic as Carrozzeria Ghia in its pomp. The Supersonic line was just one of its celebrated pieces of work. This is all the more remarkable given that this series of cars was created during a period of growth laced with constant uncertainty. That this firm even existed during the early 1950s was remarkable given that its factory had been razed during the Second World War.

Considered an enigma by motoring historians, the firm's founder and guiding light, Giacinto Ghia, wasn't lacking in ambition, even if the specifics of his early life are shrouded in conjecture. Born on 18 September 1887, this bootstrap entrepreneur served an apprenticeship at a small firm of horsedrawn carriage builders while barely in his teens, before concluding that the new-fangled automobile was far more exciting. Turin was at the centre of Italy's nascent motor industry, and he found employment with several manufacturers, over time progressing from humble mechanic to experienced test driver. After spells with STAR (Società Torinese Automobili Rapid) and Diatto, and following a stop-start competition career that was rewarded with little success, Giacinto Ghia found himself unemployed in 1915 after he sustained serious injuries in a car accident. The Diatto he was testing overturned and he broke both his legs.

The crash turned out to be life-changing on many levels. While physically he never entirely recovered, he did return to motor sport, competing twice on the Targa Florio, among other events. However, Ghia reverted to his coachbuilding roots after Diatto contracted him to construct wooden bucks ('*socca*' in Italian) on which car bodies could be fashioned. He rented a small workshop on Via Pettiti and set about establishing a name for himself. He did so with assistance from a shadowy figure known only as *Signor* Gariglio. Little is known about him, but it is widely held that he provided the financial backing. Anecdotal evidence suggests he may have been a butcher. Ghia & Gariglio expanded rapidly, moving to a much larger factory in 1921, and the firm landed a contract to supply bodies for Fiat's dashing 501S model. A year later, it began fashioning landaulette outlines for the SPA Tipo 23, too.

But storm clouds were gathering. In 1924, Gariglio departed, and Ghia found a new collaborator in Giuseppe Actis. This relationship proved short-lived. The newly renamed Carrozzeria Ghia & Actis lasted only two years

● While the gentleman in the foreground admires the latest Chrysler 'Hemi' V8, the ladies in the background are rather more taken with the Ghia-crafted, Exner-penned K-310 show car
Giles Chapman Library.

Jaguar XK 120 Supersonic

The coachbuilder: the birth, decline and rebirth of Ghia

before the partners split. At this juncture, Ghia employed a staff of around 30, and his clientele now included royalty. He continued to clothe assorted Fiat models but in addition added marque names such as Lancia, Itala, Isotta-Fraschini, Alfa Romeo and Bugatti to the company's repertoire. He also foretold the firm's future alliance with Chrysler Corporation after bodying a Chrysler 75 as a one-off commission.

By the dawn of the following decade, Giacinto Ghia had cemented his reputation as one of the premier coachbuilders of the era. According to *Motor Italia* magazine in February 1930: 'If you want not just a good, safe automobile, but a beautiful car, consult a true artist.

'As the great couturiers have to study closely with clothes, and which colours best suit different types of woman, so the artistic coachbuilders must interpret every model of chassis in a different manner. Only then will the finished car converse and enhance the fundamental characteristics – that is to say, the personality – of the model. An ordinary coachbuilder can make an accurate and diligent job with all the toil and patience of a worthy artisan, but just compare the same chassis clothed by a great coachbuilder. The result and most successful creations of Carrozzeria Ghia of Turin …readily show the class of our most distinguished coachbuilders.

In 1930, Ghia moved to an even larger facility at Via Tomasso Grossi 8, where the firm manufactured the Fiat 514 Coppa delle Alpi and 508 Coppa d'Oro sports car bodies, in addition to more prosaic saloon and coupé variants. Later that decade, Ghia began an alliance with two men who made a great impact on the company's fortunes, albeit for very different reasons.

Former Stabilimenti Farina and Pinin Farina alumnus Felice Mario Boano went it alone and established his own firm building wooden body bucks. Ghia – who began in the very same trade – became one of his first clients, and over time their relationship blossomed. He recalled in David Burgess-Wise's book *Ghia: Ford's Carrozzeria*: 'Ghia was a good friend, and a great worker, but not much of an administrator. I had to help him sometimes'.

Ghia's other key ally was a man whose name has largely been forgotten. Former motorcycle racer Count Mario Revelli di Beaumont was a pen-for-hire. He was a stylist from Rome who, over four decades, worked closely with coachbuilders such as Pinin Farina, Bertone and Viotti, plus mainstream giants like Simca and General Motors. In addition to a brilliant designer, he was also a seasoned inventor, with more than 30 patents to his name. The nobleman devised the adjustable steering column and central locking among other automotive features we nowadays take for granted. It was a line of study perhaps inherited from his father, Abiel Bethel Revelli di Beaumont,

● Pedestrian-looking Fiat 1500 rebody was as restrained as the Supersonic line was flamboyant. Ghia soon became renowned for its daring designs *Giles Chapman Library.*

who patented the Villar Perosa, the world's first submachine gun to see active service.

Count Mario was evangelical about streamlining, and he collaborated with Ghia on the wind-cheating Fiat 508-based Mille Miglia coupé that caused a furore when unveiled in 1933. His outlines for Fiat 1500 and Lancia Augusta and Aprilia platforms were as audacious as they were striking, while the Revelli di Beaumont-styled Alfa Romeo 6C 2500s produced in 1939-40 would likely have attracted even greater hoopla on the world stage had Europe not been plunged into the hell of conflict.

Automobile manufacture in Italy didn't end entirely during the Second World War, but production was severely limited. Coachbuilders were co-opted into working for the armed forces, with the likes of arch-rival Viotti producing military-specification Lancias until its factory was flattened by bombing in 1941. Bertone and Pinin Farina, meanwhile, bodied ambulances, while many smaller concerns made everything from aircraft panels to field kitchens. No great fan of the fascists in control of his country,

Giacinto Ghia nevertheless got off relatively lightly, at least to begin with. His firm's contribution to the war effort was the production of trailers for the Italian army, although it also made bicycles that were sold on the black market to transport-starved locals.

Allied bombing raids increased in frequency and ferocity during 1943-45, when many workshops fell under the control of the German military authorities. Turin's industrial heartland was largely levelled as the Allies attempted to halt supplies to enemy armies by systematically targeting any factory deemed to be contributing to the war effort. Ghia's facility was bombed seven times before a direct hit all but demolished it in late 1943. The company founder, who had departed for the hills around Turin as the city was evacuated, returned to the site dejected but determined to rebuild. He was supervising this reconstruction when he died of heart failure on 21 February 1944. He was just 56.

With the talismanic principal now gone, Ghia's widow attempted to right the ship, only to pass away soon afterwards herself. Then Boano and Giacinto

● The one-off Plymouth XX-500 marked the start of a long and mutually beneficial relationship between Carrozzeria Ghia and the Chrysler Corporation *Giles Chapman Library.*

The coachbuilder: the birth, decline and rebirth of Ghia

● Dodge Firearrow was one of several showstoppers created by Ghia and styled by Chrysler's styling chief Virgil Exner. It was reputedly good for 135mph in 1954 *Giles Chapman Library.*

Ghia's brother-in-law and company general manager Giorgio Alberti stepped into the breach as battles raged between other family members over who inherited what. Alberti hoped to assume control, but Boano bought out the various parties to take over the stricken *carrozzeria*. Having his own workshop and body-making facility helped smooth the transition as construction began on a new home for Ghia close to the smouldering ruins of the previous one.

The latest factory opened in 1946, with the firm operating under the name of Carrozzeria G. Ghia Successori SPA. *Motor Italia* magazine was quick to trumpet the firm's renaissance and stated: '[It is] ready to resume its battle station in that peaceful contest where the habitability and good sense of Torinese coachwork will quickly be triumphant again'. While some substance was lost in translation, this gushing prose was more than hyperbole. Under Boano's stewardship, the firm picked up briskly from where it had left off before the war, building ever more *outré* one-offs and small-series offerings.

A gifted designer himself, Boano nevertheless benefited greatly from having illustrator and Pinin Farina alumnus Fedele Bianco on the payroll. Another name largely airbrushed out of coachbuilding history, Bianco conjured a raft of spectacular outlines that were applied to all manner of platforms, spanning Fiat and Talbot-Lago, Lancia and Delahaye. Such coachwork invariably featured spats that concealed all four wheels. Dubbed 'Flamboyants' by the motoring media of the day, these wildly over-bodied creations were darlings of the *concours d'elégance* circuit. Alongside, Ghia also produced relatively affordable, more soberly-styled '*fuoriseries*' models based on Fiat 1100/1500, Alfa Romeo 6C 2500 and Lancia Aprilia chassis. Most were styled by Boano, but Giovanni Michelotti also contributed his artistry as a freelance stylist.

In 1947 Alberti departed during the reorganisation process, with Boano, his young son Gian Paolo, and financial officer Luigi Sabille remaining as the major shareholders. The big news, however, was the arrival of Luigi 'Gigi' Segre as sales manager. He made an instant impression, delivering new clients to the *carrozzeria*, and was soon made a partner in the business. The following year saw the establishment of a subsidiary, Carrosserie Ghia-Aigle, which was initially based in the small town of Aigle, Switzerland. Boano joined forces there with Pierre-Paul Filippi, also from Turin, in the hope of drumming up further business outside Italy. However, despite the firm producing a tranche of one-offs and ultra-limited production runs on MG, Jaguar, Delahaye, Lotus, Alfa Romeo and Porsche platforms, the venture was not a great success. By the end of the decade, the Swiss arm had morphed into a general bodyshop. The link

● The wild DeSoto Adventurer II may have looked quintessentially American but it was designed and constructed in-house at Carrozzeria Ghia. This unique car was subsequently owned by the King of Morocco *Richard Heseltine collection*.

with Turin was severed even though the Ghia part of the nomenclature was retained until the firm closed down in the early 1980s.

That Carrozzeria Ghia saw out the 1940s was nothing short of miraculous, but the company was about to embark on a period of further expansion. During this era, there was considerable cooperation between American and Italian car manufacturers via the Marshall Plan. In a roundabout way, this exchange of information led to Segre forging a close working relationship with Chrysler Corporation vice-president C.B. Thomas, who handled his firm's export sales.

In 1950, Ghia received a Plymouth chassis from Detroit, the idea being that the Turin styling house would use it to build a unique car that could function as a calling card for drumming up new business. Chrysler, for its part, was keen to learn more about the world-renowned 'Italian Line' and the methodology of how coachbuilders went about creating one-offs at such a lightning pace. Opinion-forming 'dream cars' – 'concept cars' in today's parlance – were still a novel marketing weapon in the car industry's armoury, and Chrysler was eager to be in the vanguard of the trend. It wanted to quash its dowdy image and impress on the world that it was a forward-looking manufacturer with a keen eye on design. The resultant XX-500 prototype was not particularly attractive, and in many ways

was derivative of an earlier Ghia-built Alfa Romeo 6C 2500. Clearly it did the trick. Chrysler had canvassed other coachbuilders, Pinin Farina included, but was sufficiently impressed to make Ghia its dream car-building partner.

The long-forgotten XX-500 became the first of many official Chrysler 'Idea Cars', the influential K-310 among them. This 1952 super-coupé was penned in its entirety by Chrysler's Advanced Styling Group principal Virgil Exner, with Ghia acting as subcontractor. Nevertheless, construction of this car marked the start of a symbiotic relationship, with ideas continuously flowing between Turin and Detroit. Umpteen show queens followed, some styled internally by Chrysler, others by Ghia. *Torino Motori* magazine commented in period:

> The Ghia line of these cars brought a note of gracefulness to these elephantine chassis, softening the intersection between the various elements of the car – wings, body, cockpit – and reducing the mass of chromed components which characterise American cars of all periods (and which are of questionable aesthetic value) to rational proportions.

There were, however, some serious problems at management level within Ghia. Boano and Segre

The coachbuilder: the birth, decline and rebirth of Ghia

● Left: Ghia enjoyed great prosperity during the 1950s, thanks in part to the construction of its open-top Jolly variants of Fiat 500s and 600s, Renaults and Lambretta trikes *Motorsport Images*. Below: Ghia was tasked with styling the elegant Fiat 2300S *Giles Chapman Library*.

clashed constantly, which hastened the former's departure from the company. With financial backing from Turin's Jewish community and Stabilimento Monviso founder, Allesandro Casalis, Segre was able to acquire shares held by his partners and finally assume complete control in 1953. It was at this juncture that Segre began working with Giovanni Savonuzzi, which heralded many of Ghia's more memorable creations of the 1950s. In addition to the Supersonic line, the *carrozzeria* continued to fashion one-offs and flights of fantasy for the beautiful people, including a brace of reworked Rolls-Royce Silver Wraiths, a long-wheelbase, four-door Cadillac convertible for the King of Saudi Arabia, and several Ferraris. Ghia was also involved in prototyping what became known as the Karmann-Ghia, the authorship of its outline being a source of dispute among historians ever since.

It was abundantly clear that Ghia lacked production capacity. It missed out on the opportunity to build cars in volume for the likes of Lancia and Alfa Romeo due to severely restricted factory space, so in 1953 the company entered a joint-venture with Stablimento Monviso, which had the requisite space and capability. It initially produced a series of *elaborata* versions of production Fiat cars under the Ghia-Serie Speciali banner, these having fancier, customised trim than standard offerings but not being coachbuilt in the accepted sense. Three years later, the firm was entirely absorbed by Ghia. Monviso products, including its range of Fiat 600 Multipla-based commercial vehicles, now wore the Ghia nameplate.

At around this time, Segre apparently had another brainwave, and began selling Fiat Jolly convertibles – pared-back 'beach cars' that rapidly became a must-have for the rich and famous. No yacht of any consequence was quite complete without a Ghia Jolly waiting for its occupants on the quayside, or sometimes even travelling the seas aboard the vessel itself. The Jolly is said to have been the brainchild of playboy industrialist Gianni Agnelli, head of Fiat; according to legend, he wanted a small car that would serve as a land tender but also fit on the back of his 82ft ketch *Agneta* as he cruised the Mediterranean. Carrozzeria Ghia was asked to build it and the prototype was completed in time for the 1957 Turin motor show; rival *carrozzerie* Frua and Francis

● Ghia-bodied Shelby Cobra 427 was made shortly after Alessandro de Tomaso assumed control of the Turin coachbuilder *Motorsport Images.*

The coachbuilder: the birth, decline and rebirth of Ghia

● Ghia as a marque: the Ghia L6.4, left, was patronised by Hollywood's 'Rat Pack' and US presidents alike, albeit only briefly *Giles Chapman Library*. The Ghia 450SS, below, was rooted in a Fiat proposal, but emerged based on Plymouth Barracuda foundations *Giles Chapman Library*.

● The Tom Tjaarda-styled Mustela 2+2 from 1970, below, was one of numerous prototypes initiated by Ghia's then principal Alessandro de Tomaso. As with so many other of his schemes, he quickly lost interest in it
Giles Chapman Library.

Lombardi also displayed cars built along similar lines, but Ghia's offering was unique in still being recognisably a Nuova 500.

Whether or not Agnelli ever owned the actual show car, or even personally commissioned its construction, is debatable. There is no proof to suggest he did. What's more, this was always going to be a production vehicle rather than a playful one-off. Segre was thinking big by thinking small. Nor, strictly speaking, was it intended for use only as a yacht tender. Ghia's brochure from the time described the 'Jolly de Plage' (roughly translating as 'Joker of the Beach') as being equally at home on the golf course or on hunting expeditions! A newly purchased factory on Turin's Via Agostino de Montefeltro was soon slicing open dozens of baby Fiats, adding a 600-based Jolly to the line-up in 1958.

Notable customers included shipping tycoon Aristotle Onassis and US President Lyndon Johnson, who used his on his Texas ranch, and not forgetting Hollywood A-listers Grace Kelly, Mae West, Yul Brynner and John Wayne. Ghia then developed the theme by offering a Jolly version of the Renault 4, and also a three-wheeled take on the popular Lambretta scooter aimed at the Far East market.

The new plant had been acquired from Carrozzeria Frua in 1957, its principal Pietro Frua being installed as Ghia's chief designer following Savonuzzi's departure. It was a short and stormy engagement as fissures in his relationship with Segre appeared almost immediately, in part over development of the Renault Floride production car (and who should be credited with styling it). Following a scrappy divorce and equally spirit-sapping lawsuit, Frua secured legal recognition and went back to running his own show, but Ghia retained his factory. However, small runs of Imperial limousines for Chrysler, plus the occasional show car, wasn't sufficient to keep workers occupied, which prompted the decision to market cars directly under the Ghia nameplate. This began with a Fiat 1100-based coupé and estate car. The name was then applied to the ill-starred Dual-Ghia based on the Chrysler Firebomb show car and sold exclusively by American trucking magnate Eugene Casaroll. This glamorous machine was adopted by the likes of Ol' Blue Eyes himself, Frank Sinatra (who had two), Dean Martin and Ronald Reagan who, legend has it, lost his to Lyndon Johnson in a game of poker. The unrelated Ghia L6.4 was similarly aimed at American glitterati, but only 26 were made.

Segre remained convinced Ghia needed to chase volume production. In 1959, he joined forces with typewriter manufacturer Arrigo Olivetti and car accessory firm Fergat to form Ghia offshoot OSI (Officine Stampaggi Industriali). This new firm would build niche vehicles for mainstream producers starting with the Innocenti 950S Spider which was styled by first-time designer, Tom Tjaarda. It then won the contract to shape and build the Fiat 2300S Coupé. But dark clouds were gathering over Ghia. Chrysler, the firm's main benefactor for more than a decade, continued to bankroll construction of

Jaguar XK 120 Supersonic

The coachbuilder: the birth, decline and rebirth of Ghia

● The beautiful Tom Tjaarda-penned De Tomaso Pantera remains one of the best-selling supercars of all time; it cost its backer – the Ford Motor Company – a fortune in warranty claims, however *Richard Heseltine collection.*

Crown Imperial limousines plus a batch of 50 gas turbine-powered evaluation prototypes. However, following Segre's sudden death in 1963, commissions from Detroit evaporated amid rumours of 'creative accounting' from the Italian side. Similarly, hopes of the lovely C.230 prototype replacing the Fiat 2300S were dashed, although the outline was subsequently reworked to form the basis for the Plymouth Barracuda-based Ghia 450SS. It similarly failed to find favour. The promising Ghia 1500GT, with its Gilco chassis and four-cylinder Fiat power, wasn't successful either.

With Segre gone, Giacomo Gasporado Moro was engaged as administrator for the floundering business. OSI was hived off, and Ghia was sold to Rafael Leónidas Trujillo Martínez, the playboy son of assassinated Dominican Republic dictator Rafael Leónidas Trujillo. A respected businessman, Gino Rovere, was hired as managing director but the firm's slide continued unabated. Former Maserati and Bugatti racer Rovere himself passed away in 1964, and the gadfly new owner apparently had zero interest in running a business of any description. So much so, Trujillo Martinez's sole demand as owner of Ghia was to request custom-designed labels for the wine bottled at his vineyard.

Ghia continued to lose business, although there were one or two bright spots. Moro had managed to lure Giorgetto Giugiaro away from Bertone, and the brilliant young designer styled cars such as the Isuzu 117 Sport and Maserati Ghibli for his new employer. Giugiaro recalled: "There were two reasons for joining Ghia. One was economical. I was married, we had just had our first child, and Ghia was offering me a management position. That represented a step forward, but it was also a way of testing myself; showing what I was capable of personally and professionally without Bertone behind me."

His stay was shortlived, however. In 1967, Trujillo 'sold' Ghia to Allesandro (formerly Alejandro) de Tomaso, the Argentinian émigré, so-so racing driver, and car builder. Earlier in the decade, he had made a series of track-orientated machines including Formula Junior and Grand Prix single-seaters, and

● The Ghia Coins show car from 1974 boasted only one door, which was sited at the rear. Stylist Tom Tjaarda recalled it as a "rush job" and it was the car he was least proud of designing *Richard Heseltine collection.*

34   Coachbuilt Cars

Jaguar XK 120 Supersonic

35

The coachbuilder: the birth, decline and rebirth of Ghia

● Inside the working Ghia viewing studio c1975, right to left, Ford Flashback show car, Urban Car concept and an unidentified styling model possibly from the Mustang II programme *Giles Chapman Library*.

also engaged Ghia to build the pretty Vallelunga road car. Trujillo had been imprisoned and needed money to hasten his release. De Tomaso handed over the necessary funds and in doing so assumed control of the styling house. Strictly speaking, the purchase was financed by his American brother-in-law Amory Haskell of Rowan Controller Industries, but de Tomaso had complete day-to-day control. His subsequent wrecking ball approach to man management was not appreciated by either Moro nor Giugiaro.

"I first met de Tomaso in 1958 or 1959," Giugiaro recalled.

> I could tell numerous stories, but one will suffice. During that period, he kept trying to persuade me to convince Bertone to develop his sports car. Years later, when I joined Ghia, de Tomaso was already producing the Vallelunga with a Ghia body. I designed the Mangusta for him, as well as the Pampero and a *barchetta* for circuit racing. The managing director at the time, Moro, told me there were rumours de Tomaso might become the major stockholder of Ghia. When this happened,

Moro was fired, and I did all I could to reduce the length of my contract from three years to a single year. In 1967, I was able to leave Ghia and set up my own firm.

Nevertheless, Giugiaro continued to contribute designs for two more years as a subcontractor even if some, such as a proposal for a new Checker taxicab for New York, were farmed out to fellow designer Tom Tjaarda. This likeable American became the *de facto* principal stylist in 1969 after de Tomaso persuaded him to leave Pininfarina. He was immediately tasked with styling two prospective Lancia show cars, the Fulvia 1600HF Competizione and the Marica. De Tomaso, meanwhile, had already sunk his hooks into Ford and was hoping to persuade the firm to acquire the ailing Italian brand. This scheme failed, but Ford did sponsor a new de Tomaso supercar – the Pantera. Unveiled at the 1970 New York Auto Show, and sold in the USA via selected Lincoln-Mercury dealers, warranty claims soon began to mount up until Ford ended its involvement in 1972. Unbowed, de Tomaso continued to sell variations of the Pantera in Europe as late as 1993.

De Tomaso's entrepreneurial – perhaps opportunist

● The Ghia studio under Tom Tjaarda was responsible for penning the hugely successful first-generation Ford Fiesta, introduced in 1976. It beat off rival design proposals from elsewhere within the Ford empire *Ford Motor Co/Newspress.*

– business methods came into full force in 1969, as he managed to engineer the takeover of Carrozzeria Vignale, primarily to get his hands on its modern manufacturing facilities. Company founder Alfredo Vignale perished in a car accident just one day after the deal was done. Treading lightly through the fallout between Ford and de Tomaso over the Pantera, Tjaarda continued to work on new designs, including several projects for Dutch firm DAF, but not all of them made the transition from paper to reality.

He recalled to the author: "Allesandro was always looking towards the next project, the next deal, so he just forgot about things and moved on. He would have 10 ideas on the go at any one time and of those maybe one would become reality. He was incredibly volatile and would fire you on the spot if he didn't like something. He would then phone you the following day and want to know why you weren't at your desk. That was just him. You either got used to it and learned to use it to your advantage, or you quit."

Having already invested heavily in Ghia, Ford acquired de Tomaso's remaining 16 per cent minority shareholding on 8 January 1973. A month later, it was merged with the Bruino-based Italian Design Studio (IDS) that had worked extensively with Ford of Europe to form Ghia Operations. The deal saw IDS styling chief Filippo Sapino return to Ghia, his first employer 11 years earlier, as head of the design department under Tjaarda, who was appointed studio chief.

Ghia was no longer a coachbuilder, and nor did it act as a design consultancy for outside customers. It became an advanced design department of Ford, producing a yearly array of concept cars to foretell future styling trends as well as shaping Ford production cars such as the Tjaarda-designed, first-generation Fiesta.

The Ghia badge was also used to denote top-of-the-range Ford production models in Europe, the USA and South America. In many ways, this branding exercise sullied Ghia's once-proud standing as one of the great names in Italian coachbuilding. It became mere shorthand for higher trim levels on everything from a Ford Escort to a Mercury Monarch. The Ghia nameplate was finally retired in 2010, the Turin studio having closed its doors nine years earlier, and its passing received scant media coverage.

While Ghia has been consigned to history, landmark creations such as the Supersonic ensure its glory years will never be forgotten.

Chapter Three
The designer: the complex genius of Giovanni Savonuzzi

He had one foot in the times and the other in a world of his own. Giovanni Savonuzzi was a man apart; a maverick who lived on his own, precedent-upsetting terms. Had he only the Supersonic to his name, this defiantly self-directed engineer and educator would be – should be – assured at least demi-god status. However, only in retrospect have his many achievements been trumpeted, and trumpeted loudly. Visionary designer and sometime colleague Dante Giacosa described Savonuzzi as being: 'Intelligent and creative, full of drive, and a tireless worker'. Nevertheless, this brilliant engineer, artist and aerodynamicist was not a 'name above the title' star in the automotive arena. He wasn't a fame-chasing self-publicist, nor was he often given to reminiscing in his autumn years.

It was generally left to Savonuzzi's paymasters to rake in the plaudits. As historian Piero Casucci noted during a rare interview with Savonuzzi for *Grandi Automobili* magazine in 1985:

> He rarely feels disposed to speak of himself, as if what he has achieved is the most natural thing in the world. He thanks only God and humanity for having been able to express himself freely at the service of all men, pursuing just one goal: to materialise in some way, the supreme need to make and create.

Savonuzzi's backstory, therefore, is one where myth and reality sometimes overlap. Deciphering the actual and the apocryphal is no easy task, but one thing is abundantly clear: his influence is still being felt in the car design world. Born in Ferrara in northern Italy on 28 January 1911, he displayed an early interest in engineering and in 1939 graduated with a degree in Industrial Mechanical Engineering from the Polytechnic of Turin. Following a brief spell working at Fiat Aviazione, the car manufacturer's aeronautics arm, he joined the Italian army. As Europe descended into the turmoil of the Second World War, he distinguished himself while serving in Albania, rising through the ranks to become a captain. Savonuzzi later served as the leader of a partisan resistance group, working hand-in-glove with American liberation forces, and was garlanded for helping secure the Italian peninsula for the Allies.

While industries and infrastructure lay shattered from the Alps to Sicily, Italy welcomed a new manufacturer of sports and racing cars – Cisitalia. Savonuzzi joined this start-up operation a year later, his presence being felt from the outset. Marque instigator Piero Dusio was nothing if not ambitious. Even now, the grandiosity of his vision seems breathtaking. This fascinating character was born in Scurzolengo, south-west of Turin, in October 1899. A natural sportsman, his footballing career with Juventus was curtailed by a knee injury, but he found a perfect substitute in motor sport. A savvy businessman, Dusio earned fortunes in real estate and textiles, which paid for his racing exploits. A *pilota* gentleman, he was sufficiently talented to place sixth overall in the 1936 Italian Grand Prix at Monza aboard a Maserati 6C. Becoming a manufacturer in his own right represented a natural step, somewhat in the Enzo Ferrari idiom. Dusio established Consorzio Industriale

● A free-thinking visionary, but humble to the last, Giovanni Savonuzzi designed some of the most instantly recognisable cars ever made, but he was not one to crow about his achievements
Alberta Savunuzzi, with thanks to Peter Vack.

The designer: the complex genius of Giovanni Savonuzzi

Sportive Italia (Cisitalia) in 1944

By his own admission, he was no engineer, but he did have a knack of recognising and enabling burgeoning talent. The open-wheel D46, the model that established the marque trackside, was largely the work of Dante Giacosa. Nevertheless, Giacosa stated in his 1979 book *My 40 Years With Fiat*: '[Savonuzzi] played a decisive part in perfecting the single-seater prototypes'. Powered by a tiny 1.1-litre four-cylinder engine, this skimpy device punched above its weight, with Tazio Nuvolari driving one to victory in the Coppa Brezzi in September 1946.

This was, however, merely the opening salvo. Savonuzzi played midwife to two landmark models: the time-defying 202 coupé, and the 202CMM, the latter better known by its 'Nuvolari Spider' alias. Nuvolari, nicknamed the 'Flyin' Mantuan', almost pulled off a disruptive win on the 1947 Mille Miglia in this be-winged roadster. The veteran race ace wasn't in the best of health, yet somehow his Cisitalia was eight minutes clear of the chasing pack at the half-way point in Rome. His progress was all the more remarkable given the car packed a mere 1098cc and maybe 65bhp.

The first post-war running of the great race was held in June that year but there was nothing at all summery about the weather. Nuvolari and co-driver Francesco Carena battled monsoon conditions on the closed *autostrada* between Turin and Milan, only for the Cisitalia's ignition system to take in water. Valuable time was then lost while the distributor dried out. The lead may have changed, but Nuvolari wasn't done yet. Once up and running, he pressed on as only he knew how, but ultimately had to settle for second place behind Alfa Romeo 8C 2900B duo Clemente Biondetti and Emilio Romano. Class honours were some consolation, at least.

Cisitalia was stopped short of an upset victory, but it had more than made its mark. If Nuvolari's performance wasn't enough to guarantee banner headlines, Cisitalias also finished third and fourth overall. This was a remarkable achievement for an operation that had constructed its first car barely a year earlier.

It is widely held that Cisitalia made as many as 30 202MM roadsters, but its fame was fleeting when

● *Cisitalia virtually owned the hugely popular 1100cc class in early post-war Italian racing with its Fiat Millecento-powered single-seaters Richard Heseltine collection.*

● Right: Piero Taruffi driving a Cisitalia D46 to victory in the Circuito di Caracalla race in Rome in June 1947 *Motorsport Images*. Below: Cisitalia founder Piero Dusio flanked on the left by Piero Taruffi and on the right by brilliant designer-engineer Giovanni Savonuzzi *Alberta Savonuzzi*.

The designer: the complex genius of Giovanni Savonuzzi

compared with that of the 202SC coupé, and that's even though Savonuzzi's role with it has often been under-reported in print.

Given that most sports cars of the day were reheated pre-war models – often with cycle-wings and square-rigged bodies – the fixed-head Cisitalia was breathtakingly advanced. It offered a fully-enveloped body and a beautifully arched roofline, and yet it stood barely shoulder-high. Among the delightful details was its wide oval grille, which remarkably was cast in one piece, slats and all. Numerous coachbuilders were approached with a view to transforming Savonuzzi's sketches into full-scale reality, and ultimately Pinin Farina was chosen. This landmark design was positively jaw-dropping in 1947, and arbiters of beauty have been hailing it a masterpiece ever since, and an early example of a 'GT' as they came to be recognised among roadgoing cars.

Savonuzzi not only penned the car's outline, he also devised the mechanical layout, and even acted as test driver. Variations on the theme influenced designers on both sides of the Atlantic. For example, the fiendishly complex 202 SMM Aerodinamico Coupé – or 'Savonuzzi Coupé' - featured huge tailfins that were well ahead of the styling fad that gripped Detroit throughout the following decade. The Cisitalia's original fins, though, weren't there as styling tinsel; they were positioned to aid stability at high speed. It fell to Alfredo Vignale to translate this technically

● Left: general astonishment at the Cisitalia 202 upon its unveiling in 1947, with the towering Savonuzzi believed to be on the far-right *Motorsport Images*. Below: the 202 enjoying the limelight as a sculptural design icon in New York's Museum Of Modern Art *Giles Chapman Library.*

● The always style-conscious Henry Ford II at the wheel of a Cisitalia 202 cabriolet in 1948. He subsequently sponsored the build of some Ford V8-engined prototypes, but they didn't make the leap to series production *Richard Heseltine collection.*

advanced vision into reality after several other coachbuilders baulked at the complexity.

While the D46 earned valuable revenue, and the open and closed 202 road and competition cars caused a huge splash in the contemporary motoring media, other schemes proved anything but successful for Cisitalia. Dusio over-extended himself in building a Grand Prix challenger, and it almost ruined him. With a brains trust that included Ferdinand Porsche, Rudolf Hruska (who later engineered the Alfa Romeo Giulietta and Alfasud) and future tuning expert Carlo Abarth, the resultant single-seater – referred to as the Type 360 in Porsche lore – featured a supercharged 1493cc flat-12, mounted amidships. Unfortunately, it was scuppered by a lack of finance, with any glory being garnered in South American speed record bids rather than on the racetracks of Europe.

This Grand Prix challenger also hastened Savonuzzi's departure from Cisitalia. He had devised a 1.5-litre, double overhead-camshaft, four-cylinder engine for a variety of future road and track applications. With this work in hand, he reasoned that the firm needed to find its feet before committing time and valuable resources to building and fielding a top-flight racing car. Anecdotal evidence suggests he was also deeply unhappy that Dusio paid a large sum of money to secure the release of Ferdinand Porsche from prison in France, so he could design the car. Savonuzzi's brother Alberto had been killed by the SS in 1943, which maybe explains why he wasn't keen to collaborate with anyone tainted by the Nazi regime.

Savonuzzi's next stop was Vincenzo Leone's Officine Elettromeccaniche workshop in Turin. There in 1949 he found a collaborator and foil in former aircraft engineer and mechanic Virgilio Conrero. It was the start of an enduring relationship. Their first project under the SVA (Società Valdostana Automobili) banner was a tiny 500cc single-seater built at the behest of Torinese racing driver Ugo Puma. The beautiful Falcone resembled a scaled-down Grand Prix car and was powered by a Moto Guzzi 120-deg twin-cylinder engine. Unfortunately, it rarely ventured trackside, with amateur driver Puma instead turning his attention to campaigning sports cars.

The duo then followed through with a 1.5-litre single-seater that also made only occasional race

The designer: the complex genius of Giovanni Savonuzzi

● Left, Ghia's sleek, regular coachwork on the Alfa Romeo 1900 SS chassis contrasts with, right, the dainty Cisitalia 202 Savonuzzi's styling work *Motorsport Images.*

appearances. Conrero went it alone in 1951 (see next chapter) on forming Autotecnica Conrero, initially specialising in car repairs and performance tuning, while Savonuzzi then became a freelance designer and a lecturer at the Politecnico di Torino. His insatiable curiosity led him to look beyond the automotive field. During this period, Savonuzzi designed a speedboat which was piloted by Massimo Leto di Priolo to a remarkable 83mph (134kph) on Lake Idroscalo, Milan in 1953. The sometime sports car racer claimed a new world water speed record in the process. In a roundabout way, this involvement in waterborne machinery led to Savonuzzi renewing his association with Cisitalia as a consultant.

Piero Dusio had by now become embroiled in the Péron regime's bid to establish a motor industry in Argentina. He was a prime mover in the formation of the Autoar concern there, only to be elbowed out shortly thereafter. In Turin, it was left to his son Carlo to try and halt the brand's slide into oblivion, yet Cisitalia was entering its twilight years. Plans to build a car with Ford backing ultimately came to nothing despite considerable expenditure of time and money. Savonuzzi had met Henry Ford II in Paris to thrash out ideas, and styled an extremely pretty variant himself that contrasted strongly with a distinctive – if not overly attractive – Vignale-shaped alternative offering. Bold plans to equip the 202 with an adapted marine engine made by BPM (Botta & Puricelli Milano) also proved a costly distraction. The Aldo Brovarone-penned, Fiat 1100-103-based 33DF Voloradante coupé, by comparison, was a more conventional proposition. There was nothing complicated about its makeup; there was no reaching for the stars here and Savonuzzi's involvement was marginal, if he had any involvement at all. It also failed to find favour, with just four being made. In 1954, the Voloradante was quietly dropped and Cisitalia slipped into limbo.

A year earlier, Savonuzzi had joined Carrozzeria Ghia as *Direttore Tecnico*. The Turin styling house was in a state of flux despite its relative prosperity. The relationship between company principals Luigi Segre and Mario Boano had soured, with the latter departing for Fiat. The Turin workshop was by now

● The Chrysler Dart in period boasted a retractable metal hardtop that was subsequently replaced with a more conventional fabric hood. It was also known informally as the "Super Gilda" *Richard Heseltine collection.*

churning out show cars and small-series products at a dizzying rate, with stylists Giovanni Michelotti and Pietro Frua retained as freelancers, even if their contributions were rarely recognised in public. As with all coachbuilders of the time, the glory would be conferred on the business rather than a specific individual, so untangling the many knots in the Ghia narrative regarding who did what is almost impossible. There's no 'I' in team, and all that. What is beyond doubt, though, is that the Supersonic series of cars was entirely the work of Savonuzzi and Conrero together.

Some sources claim Savonuzzi was also responsible for styling a pair of Ghia-bodied Cadillac Series 62s. The first one built in 1953 at the behest of Prince Aly Khan, who gifted it to actress Rita Hayworth. Others insist that the outline design was the work of Segre. There are signature Savonuzzi flourishes that mirror the Supersonic line, especially the low roofline and taillight treatment, but it may have been just as easily a collaborative process.

The same is true of the De Soto Adventurer II which was completed a year later. Chrysler's styling chief, the legendary Virgil Exner, became firm friends with Savonuzzi during visits to Ghia while scrutinising the construction of several show cars. 'Ex', as he was universally known within the design community, was a great admirer of the Cisitalia 202 SMM Aerodinamico Coupé.

A degree of confusion surrounds the authorship of the Adventurer II, as Ghia generally acted as subcontractor to Chrysler, constructing most of its 'Idea Cars' in Italy using designs from its Advance Design Studio in Detroit. Savonuzzi is sometimes credited with shaping this car, again possibly in conjunction with Segre. The styling is undeniably reminiscent of themes explored beforehand on the Supersonic, not least the gently arched beltline which terminates in circular 'afterburner' rear light clusters. Equally, it exhibited many typically Exner styling cues, such as its retractable roof/rear window. Regardless of who penned it, the car was a major hit on the show circuit before it was acquired by the King of Morocco in 1956.

The designer: the complex genius of Giovanni Savonuzzi

According to *Virgil Exner – Visioneer*, Peter Grist's in-depth biography of the American legend, Savonuzzi, Segre and Boano each claimed to have styled what in time became the Volkswagen Karmann-Ghia. You could, however, argue that the car's outline was plagiarised from the 1953 Chrysler D'Elegance show car, styled solely by Exner at his home studio in Michigan. Photos exist of his scale model to substantiate this. Tellingly, Ghia presented one of the first Karmann-Ghias off the German production line to Exner as a gift, who in turn gave it to his son, future car designer (and Ghia alumnus) Virgil Jr.

In 1955, Savonuzzi produced two dazzling one-offs that continue to polarise opinion.

The extraordinary Gilda (aka Ghia-X) was honed in the wind tunnel at the Politecnico di Torino, and followed an earlier experiment where inkblots were blown over the surface of a matching plastic model. Savonuzzi noted how the shape of the blots deformed, and designed this arrow-like projectile to mirror this. Notionally powered by a 1.5-litre OSCA four-cylinder engine, and boasting a drag coefficient of just 0.19, Ghia talked up a theoretical top speed in excess of 140mph. In reality, the car was a non-runner when displayed publicly for the first time at the Turin motor show in November 1955. Named

● The remarkable Moto Guzzi-powered Nibbio II record-breaker, left and above, was another masterwork from the pen of Savonuzzi; nibbio means kite in Italian). It was based on a Volpini chassis and had a displacement of 350cc. Count Lurani steered it to a new speed record at Monza in June 1956 (Both *Richard Heseltine collection*.

● Several designers claimed authorship of the highly successful Volkswagen Karmann-Ghia, and Savonuzzi was among their number, although he was perhaps the least vocal about it *Giles Chapman Library.*

after the 1956 film-noir Gilda starring Rita Hayworth, the car continued to make appearances at major international events right up to the end of the decade, many of them in the USA.

'Italian coachbuilders no longer confine themselves to producing elegant bodywork for luxurious large cars, or for baby Fiats,' the *Automobile Revue* annual noted. 'They now build dream cars [such as] the Gilda, exhibited by Ghia at Turin'. American magazine *Motor Trend* also made it one of the cover stars of its September 1955 issue, the strapline reading: 'X for Chrysler's Gas Turbine?' It took a further 50 years before – amazingly – it was indeed finally equipped with such a power unit. But the Gilda did inspire the Chrysler Dart show car, built by Ghia to Exner's brief in 1957. According to *Virgil Exner – Visioneer*: 'While it was being built in Italy, aerodynamicist Giovanni Savonuzzi developed the car further from his initial Chrysler-based [sic] Ghia Gilda. Extensive wind tunnel and road testing was carried out by him before it was shipped to America.' This prototype was repeatedly reworked and ultimately morphed into the Chrysler Diablo show car.

The other Savonuzzi design from the period was every bit as radical, if not quite so rapturously received as the Gilda. The final Ferrari bodied by Ghia, the 410 America more than lived up to the old American maxim of "It ain't done 'til it's overdone". Built to order of Wisconsin paper products mogul Bob Wilke, this was rather more than just a show car, with the Leaders Cards principal routinely driving his flamboyant 4.5-litre V12 device as its maker intended. It was also a regular attraction in racetrack paddocks. Wilke-backed cars claimed victory in the Indianapolis 500 on three occasions (1959, 1962 and 1968) among other major contests.

Delivered as a rolling chassis to Ghia's Turin facility in November 1954, Savonuzzi and the *carrozzeria*'s artisans set about creating a statement that mirrored Wilke's own showman's sense of style. With its large, egg-box grille, and glitzy brightwork, only the vents along the flanks hinted at the car's competition-inspired roots. Finished in salmon over anthracite, a bold chrome spear delineating the contrasting colours, the same shades were carried over into the luxuriously trimmed cabin. This bizarre machine was in no way subtle, more the sort of car that rendered onlookers moon-eyed, mouths agape, and conversing in tones fully deserving italics and multiple exclamation marks. Unveiled alongside the Gilda at

The designer: the complex genius of Giovanni Savonuzzi

the 1955 Turin Motor Show, the car was subsequently shipped to the USA where Ferrari concessionaire Luigi Chinetti borrowed it for that year's New York International Auto Show. Often referred to in later years as the 'Super Gilda', this most *outré* of Ferraris remained in Wilke's ownership until his death in 1970.

Savonuzzi departed Ghia in 1957, a year that was blighted with personal tragedy. Having already lost one sibling during the Second World War, his youngest brother Giorgio went missing following a climbing accident near Cortina d'Ampezzo. A distraught Savonuzzi headed search parties for five months, but a body was never recovered. It was against this backdrop that he departed Italy for a new life in the USA. Exner introduced Savonuzzi to Jack Cheripar at Chrysler Engineering and a job was soon in the offing. Initially interested in working in one of the firm's defence and space research departments, he instead became assistant chief engineer in the nascent Gas Turbine Research & Development wing. He remained with the smallest of Detroit's 'Big Three' until 1969, although he didn't enjoy a happy working relationship with George Huebner, the executive who created the turbine programme. From all accounts – and such accounts aren't hard to come by – Huebner was a gifted engineer and scientist but also someone who enjoyed hogging the spotlight. Colleagues like Savonuzzi were rarely, if ever, given credit for their work. The fault lines in their relationship soon began to show.

Nevertheless, the Italian *émigré* rose through the ranks, becoming chief engineer, Automotive Research in 1962. According to Peter Vack's extensive 2011 feature on Savonuzzi for VeloceToday.com, he filed more than 30 individual patents while employed at Chrysler. He also collaborated with Ghia on the construction of around 50 gas turbine-powered Chrysler working prototypes, cars that were famously evaluated by specially selected members of the public across the USA, ahead of a proposed production run. Ex-Lincoln man Elwood Engel is widely credited with styling the Turbine cars, but Savonuzzi and Carrozzeria Ghia made contributions, even if their input wasn't necessarily appreciated or requested. Savonuzzi also had the use of one of these prototypes

● The undoubted star of the 1955 Turin motor show, the Gilda was named after Rita Hayworth's titular role in the film of the same name. The power unit was an OSCA four-cylinder, although there was a pretence that it was gas turbine-propelled... *Giles Chapman Library.*

● The Chrysler Turbine was engineered by Savonuzzi and boasted styling by Elwood Engel; a run of 50 working prototypes was produced for the American company by Ghia in Italy *Chrysler/Newspress.*

The designer: the complex genius of Giovanni Savonuzzi

● Savonuzzi had long been fascinated by gas turbine powerplants like this in the roadgoing Turbine, an interest that came into full bloom after Chrysler's design czar Virgil Exner lured him away from Italy to work in Detroit *Giles Chapman Library*.

as a daily driver. Distinct from its stablemates, which were all finished in a striking bronze hue, his Turbine was given a special livery for its use in the film *The Lively Set*, which rendered it even more conspicuous.

In 1963, overtures were made to Savonuzzi to return to his homeland and assume the role of Ghia principal, following the sudden death of Segre. He turned the offer down and instead remained at Chrysler. Nevertheless, he felt increasingly marginalised by Huebner, who reputedly threatened to annul his pension should he take up a position elsewhere. At the end of the decade, Savonuzzi threw in the towel and returned to Italy anyway. According to Vack's authoritative article:

> There is good reason to believe that after he finally returned to Fiat in 1969, Huebner methodically destroyed all evidence of Savonuzzi's work, effectively erasing him from the history of the Chrysler Turbine. Archivists at Chrysler have been unable to find any trace of Giovanni Savonuzzi… [He] turned to Giovanni Agnelli at Fiat, who gave him the position of director of Research and Development, and more importantly, made up for the loss of pensions and benefits Savonuzzi was sure to lose from his departure from Chrysler… True to his threat, Huebner reduced his monthly pension to a mere $27.00.

Savonuzzi worked on numerous prototypes at Fiat up to his retirement in 1977, some of them featuring alternative power sources (he always declared, though, that hyper-efficient petrol-engined cars made the most sense). Not that he ever fully retired, at least not in the dictionary sense of the word. When he wasn't indulging his other great passion – sailing – this remarkable all-rounder continued to act as a consultant to car manufacturers while also becoming a guest lecturer at the Politecnico di Torino. He died on 18 February 1987, aged 77.

Savonuzzi's name is not always uttered with the sort of reverence reserved for those who followed in his wake, such as Giorgetto Giugiaro and Marcello Gandini, but his contribution to post-war automotive design and engineering remains immense. Here was a man capable of shaping the most starry-eyed flight of fancy and also choosing and refining real-life mechanical underpinnings. This sets him apart from most of his contemporaries. The pity is that his praises weren't sung as highly as they might have been during his lifetime.

● Savonuzzi regularly drove the sole Chrysler Turbine Car not finished in bronze. It was specially liveried for use in the film *The Lively Set*, as seen in a scene on-set, here being chased by a Ghia L6.4 *Universal Pictures*.

Chapter Four
Under the bonnet: the work of tuning ace Virgilio Conrero

He was nothing if not a self-starter. Virgilio Conrero was variously an engineer, designer and race team principal, but he is best remembered for his ability to extract improbable amounts of horsepower from even the smallest of engines. The Torinese tuning legend was known as *Il Mago* – The Magician – for good reason. Nevertheless, for all his achievements, his fame didn't stretch much beyond Continental Europe during his lifetime.

Born on 1 January 1918, Conrero's future was, to some extent, preordained. His father owned a major machine tools and munitions business, and employed more than 50 workers in his Turin works. Family wealth ensured the young Conrero wanted for nothing, but this prosperity came to a juddering halt after an explosion saw the factory destroyed. His father was killed in the enormous blast, and the business never recovered. The Conreros were soon declared bankrupt.

Young Virgilio left school aged 14 and took a job as a machinist. A year later, he began attending evening classes in the hope of becoming an aircraft mechanic. Despite regular 12-hour shifts in his day job, he did well enough in his studies to graduate top of his class. Fiat promptly offered him a job in its aircraft division. There, Conrero worked on civilian and military projects, primarily in a prototyping and development role before the Second World War intervened. In 1940, he joined the Regia Aeronautica Italiana (Italian Royal Air Force) as a mechanic, before returning to Fiat to see out the rest of the conflict working on new aircraft engines. In peacetime, the Italian aircraft industry was effectively nullified by the victors, so Conrero was reduced to repairing bicycles before he was offered a job as a truck mechanic. Having shown no previous interest in roadgoing vehicles, the future tuning colossus learned the ropes working on large diesel engines.

In 1948, he began operating out of the Officine Elettromeccaniche workshop in Turin as a machinist. It was here that he teamed up with Giovanni Savonuzzi on numerous projects, some of them under the SVA banner. The most ambitious of these was a single-seater racing car equipped with a Shorrock-supercharged 820cc four-cylinder engine and a De Dion rear suspension arrangement. Sadly, it died on the tool-room floor due to a lack of funds.

Conrero went it alone in 1951. At first he spent his time straightening out crashed road cars, but thanks to recommendations from Savonuzzi and Piero Dusio, he was soon preparing Cisitalias and other sports cars for circuit use. One of his early customers was Robert Fehlmann, who charged Autotechnica Conrero with preparing his Cisitalia for the 1952 Mille Miglia road race. The Swiss amateur driver came home seventh in the hotly contested 1.1-litre class, which led him to commission a bespoke one-off for a tilt at the 1953 running. That car was the Savonuzzi-designed, Alfa Romeo 1900-engined Supersonic.

From his minuscule workshop in 32 Via Mon Basilio, Conrero then set about building a *barchetta* under his own name. The first Alfa Romeo 1900-based Conrero sports-racer featured a bespoke twin-plug cylinder head, the knock-on effect being that he was soon inundated with requests to build similar Alfa Romeo racing engines.

● The one and only Conrero Alfa Romeo prototype ahead of the 1953 running of the Mille Miglia. Its debut run ended in calamity. Conrero alumnus, designer Vilhelm Koren, claimed a stray cigarette may have had something to do with the ensuing fire but this is disputed by historians *Giles Chapman Library*.

Under the bonnet: the work of tuning ace Virgilio Conrero

Conrero), initially fielding a quartet of Alfa Romeo Giuliettas in the Italian hillclimb championship. They dominated more than one class. However, his efforts were almost for naught following a crash in a Ferrari in May of that year. The car, driven by future Scuderia Serenissima regular Gianni Balzarini, was written off following a high-speed shunt that resulted in Conrero being hospitalised for three months. He never fully recovered, having been left with debilitating leg and arm injuries. Without the appropriate medical insurance, he came close to declaring bankruptcy, but somehow Conrero managed to bounce back to construct the first of a series of sports-racing cars in conjunction with Giovanni Michelotti in 1960.

Conrero devised further tuning gear for Giuliettas that were adopted by works and privateer entrants alike, in addition to assisting in Alessandro de

● Above: Conrero wasn't content with just tuning cars – he also built several sports-racers and single-seaters, including this Formula Junior *Richard Heseltine Collection*.
Left: Robert Fehlmann at the wheel of the Conrero Alfa Romeo at the start of the 1953 Mille Miglia in Brescia. He crashed out of the race *Giles Chapman Library*.

He subsequently developed a raft of tuning parts for the cars, in addition to offering desirable disc brake conversions. In 1957 alone, Conrero-modified Alfa Romeos claimed 28 overall or class victories in Italy and France. A year later, that figure had risen to 67, and in 1959 a remarkable 84. In that season, many more wins were accrued in France where Conrero found a ready market for his demon cylinder heads and exhaust systems.

From 1958, the silver-haired tuner operated his own race team – Conrero Squadra Corse (aka Squadra

Tomaso's fledgling Formula One bid. He also continued to dabble in the construction of small-series racing cars, some of them Lancia-powered. By the dawn of the 1960s, he had moved into a slightly larger workshop in a residential street in Turin, at 18 Via Madama Christina. While he wasn't overly popular with his neighbours, not least because of the sound of race engines being revved to valve bounce at all hours, he was able to bolster his fortunes by offering aftermarket go-faster parts for imported marques such as Renault, Simca and Peugeot. Around this

- British weekly magazine *The Autocar* managed to photograph the Conrero Alfa Romeo on the eve of its Mille Miglia bid. The article was full of praise for the design, even if the name of its creator – standing on the right in Gordon Wilkins' top-right snapshot – was spelled incorrectly *Autocar*

864   The Autocar, June 26, 1953

The chassis on its first road tests. Its designer, Virgile Conrero, is at the wheel in the picture on the right.

AN ITALIAN SPECIAL . . . . continued

structed a number of racing and sports "specials." In this latest production he selected the units which he thought most suitable, the Alfa Romeo C, or Sprint type, 1900 engine, the Lancia Aurelia transmission, rear axle with its independent suspension, and steering, and the Fiat 1400 independent front suspension, mounting them on a tubular chassis.

The engine normally has an output of 95 b.h.p. but this has been raised to 130 b.h.p. at 6,400 r.p.m. The modifica-

AN ITALIAN "SPECIAL"

THE ALFA CONRERO IS A COMBINATION OF ALFA ROMEO, FIAT AND LANCIA UNITS ON A TUBULAR CHASSIS FRAME

BUILT by an Italian aircraft engineer, Virgile Conrero, of Turin, for the Swiss driver Robert Fehlmann, the Alfa Conrero sports coupé made its debut in the Mille Miglia, but not unnaturally, as it was completed only a few days beforehand, suffered from teething troubles. It is, however, likely to figure in speed events in Switzerland during the present racing season.

Signor Conrero has previously con-

For this racing coupé Ghia have produced a low-built and squat design which in some ways is reminiscent of the Alfa Romeo "Flying Saucer."

Accompanied by racing driver R. Fehlmann on the left (for whom the car has been built) and his colleague G. Vuille on the right, the designer, V. Conrero, of Turin, surveys his work. The body is by Ghia.

The Alfa Romeo Sprint 1900 engine, fed by four Del' Orto independent carburettors, gives 130 b.h.p. instead of the 95 b.h.p. in standard form.

axles.

On this very low-hung chassis is mounted a Ghia sports coupé, which is particularly low built, as the overall height is only 3ft 11in. The wheelbase is 7ft 11½in, and the track 4ft 4¾in at the front and 4ft 4in at the rear. The overall length is 13ft 11in and width 6ft 0in, but an adequate ground clearance of 6in has been maintained. The weight in running trim is about 15cwt (1,760 lb), giving a power-weight ratio of 13.6 lb per h.p. In its first tests the Alfa Conrero is said to have attained a speed of 140 m.p.h. (225 k.p.h.) at an engine speed of 6,500 r.p.m.

Under the bonnet: the work of tuning ace Virgilio Conrero

time, he also began selling Fiat-based Formula Junior engines, in addition to designing a Formula One unit of his own. It remained stillborn, although a Conrero-Alfa Romeo unit did power Syd Van der Vyer's Lotus 18, a Grand Prix car that achieved decent results at national level. He also worked closely with Triumph on a would-be Le Mans car that never raced, and sadly remained unique for all its Michelotti-drawn beauty.

Despite branching out and working with other marques, Conrero's name remained inextricably linked with Alfa Romeo. So much so, magazines such as *Auto Sprint* and *Auto Italiana* wrote breathless editorials outlining how Alfa Romeo was secretly working on a return to the Formula One World Championship with Conrero running the show. It never happened because his friend Carlo Chiti, boss of Autodelta, became the de facto works Alfa Romeo entrant in 1965, fielding the Giulia GTA in competition (the marque didn't officially return to Formula One as a constructor/entrant until 1979). Conrero felt snubbed and turned his attention to working with Honda as a consultant on various motor sport schemes, before running assorted Renaults and Alpines on-track.

Matters took an upswing in 1969 when Conrero was approached by Romano Artioli, owner of Garage 1000 Miglia, which held an Opel agency. Would he be interested in building a competition version of the new Opel GT? Conrero overcame initial scepticism and built a brace of wide-bodied GTs ahead of the 1970 Targa Florio. It marked the beginning of a long and hugely successful alliance with Opel that lasted until the mid-1990s, during which time the Conrero team claimed honours in the Italian Rally Championship, European Rally Championship and the Italian Touring Car Championship. In later years, the revered Conrero name was applied to everything from sports-prototypes to Porsche GT weaponry, before it disappeared from the racetracks for good in 2004. The team claimed to have accrued more than 4000 class and overall victories over half-a-century, its drivers including such legendary names as Lorenzo Bandini, Ludovico Scarfiotti, Ricardo and Pedro Rodríguez, Maurice Trintignant and Jochen Rindt. René Arnoux was also briefly employed by the team, albeit as a race mechanic.

Virgilio Conrero died on 6 January 1990, his legacy assured. While not nearly as well known on the global stage as arch-rival Carlo Abarth, his name still resonates in motor sport circles as one of the absolute masters of extracting the maximum from the minimum.

● The shapely Conrero-Triumph would-be Le Mans racer featured a spaceframe chassis and a twin-cam, four-cylinder engine called the Sabrina, as used in the 24-hour race in 1959-60. It remained unique, more's the pity *Giles Chapman Library.*

Chapter Five
Firebrand: Luigi 'Gigi' Segre and his American adventures

Entrepreneurial and artistically gifted, Luigi 'Gigi' Segre seized every opportunity he could to further his ambitions. During the early 1950s, he steered the beleaguered Carrozzeria Ghia towards prosperity, and helped to cement the firm's role on the global stage. Without him, the business probably would not have seen out the decade. Accordingly, he left an indelible impression on the story of Italian coachbuilding. Unfortunately, his life was also blighted by misfortune.

Born in Naples on 8 November 1919, Segre worked for his father's construction business during the 1930s before being conscripted into the Italian armed forces. Towards the end of the Second World War he acted as the liaison between the Office of Strategic Services and local partisans, and later helped repatriate Italian soldiers from France. In the immediate post-war years, he qualified as an engineer and was employed by Ford prior to joining Giorgio Ambrosini's Siata concern. It was while working for this tuning firm and small-series car builder as commercial director that he embarked on his short but moderately successful motor sport career. Segre claimed class honours on the Mille Miglia in 1949 and 1950 driving a Siata-modified Fiat 1100B alongside co-driver Gino Valenzano.

Then Ghia chief Felice Mario Boano invited Segre to join Carrozzeria Ghia in 1950. The first fruits of their collective labours was a rebodied Plymouth, named XX-500. This bulky-looking fastback sedan didn't set the world alight and was not helped by the Italian motoring media, which made some unflattering comments about it in print. However, this now long-forgotten six-seater did a great deal to promote Ghia within the corridors of the Chrysler Corporation, which manufactured Plymouth cars. The firm's inspirational styling director Virgil Exner became its loudest cheerleader. Segre, who spoke passable English, did much to forge a relationship with Exner, with bilingual Italian-American Paul Farrago acting as intermediary and Ghia's overseas agent. Boano, however, was convinced Ghia's future lay closer to home; he thought Ghia should focus on collaborations with Italian manufacturers to build small-series cars as well as styling studies and lucrative one-offs. Perhaps inevitably, the partners split in 1953 and, following further spats among shareholders, Segre assumed total control.

There followed a move to new factory premises, and also a decisive change of emphasis as Ghia chased design and prototyping work with mainstream manufacturers while also acting as Chrysler's de facto show car-building 'skunk works.' In addition, Ghia became a marque in its own right, albeit with patchy degrees of success. Such growth spurts heaped further stress on to Segre, whose hair-trigger temper had become infamous among industry insiders. Future Ghia design director Filippo Sapino recalled in *Ghia: Ford's Carrozzeria* by David Burgess-Wise: 'Segre coughed a lot, but when he was angry, he would shout so loud that he could be heard all over the plant. The walls would literally tremble!'

Michigan-born Tom Tjaarda, the design legend who enjoyed – endured, more likely – two stints at Ghia, concurred. He told the author in 2012:

● The strong-selling Volkswagen Karmann-Ghia gave the Italian coachbuilder/design house a strong presence in the USA and elsewhere. Segre claimed he styled the car, but this has been a source of debate among historians for decades *Giles Chapman Library.*

Jaguar XK 120 Supersonic

59

Firebrand: Luigi 'Gigi' Segre and his American adventures

● The Chrysler D'Elegance of 1952 was one of the Italian-American collaborations built under the entrepreneurial Segre, and was significant as the real inspiration for the Karmann-Ghia, in considerably downsized form *Giles Chapman Library*.

Jaguar XK 120 Supersonic

Firebrand: Luigi 'Gigi' Segre and his American adventures

Above: 'Gigi' Segre was business-minded and artistically gifted, and helped steer Ghia to prosperity during the 1950s and '60s. Right: Volkswagen happily traded off its Ghia connection, as in this magazine advert *both Giles Chapman Library.*

> I joined Ghia in 1959 and really knew nothing. I was fresh off the boat and thrown in at the deep end working on what became the Innocenti 950 Spider. Segre was interested in all things American and had become fascinated with the idea of building a dragster. The problem was, he didn't really understand what a dragster was. We built this streamlined thing with a canopy called the IXG. It was supposed to have a little Innocenti engine in it, but it wouldn't fit without sticking through the hood. When Segre found this out, he was apoplectic with rage. You just had to ride it out.

The early 1960s witnessed the emergence of Ghia's OSI (Officine Stampaggi Industriali) offshoot, that offered volume production of niche vehicles for mainstream manufacturers. However, the breakneck pace was starting to tell. Something had to give, and for Segre it was his health. During a transatlantic flight in February 1963, he complained of feeling ill. After a check-up at the Henry Ford Hospital in Detroit, he was told he was suffering from gall bladder trouble. He was also informed his appendix needed to be removed as soon as possible. Segre made arrangements for a friend to perform the operation on his return to Turin and continued his business trip. Five days later, on 29 February 1963, he died under anaesthetic on the operating table. He was just 43.

Segre's personal assistant Gaspardo Moro told historian David Burgess-Wise:

> [He] was an interesting, dynamic man who travelled much. He was a man who saw far. He was the best coachbuilder; the first one from Turin to do business with the United States. Thanks to the Karmann-Ghia models built for America, for many years Ghia was better known in the United States than Pinin Farina. I remember the first time I arrived at Kennedy Airport, there was a sign advertising "Karmann-Ghia Automobiles". Ghia was famous in the whole automobile field in the United States, and it was all due to Gigi Segre. He was an exceptional man.

● A rare photo (right) of the GX1 during the prototyping stage; *Richard Heseltine Collection* it later became the mass-produced Innocenti 950S (below) and so the first production car styled by future design giant Tom Tjaarda, who was then a complete unknown. *Giles Chapman Library*.

Chapter Six
At the sharp end: Felice Mario Boano, coachbuilding workhorse

One of the most august names of coachbuilding history, Felice Mario Boano has nonetheless been ill-served by history. After completing an apprenticeship at Stabilimenti Industriali Farina, this gifted designer rose to the position of technical director before going it alone. Unfortunately, his decision to establish his own business coincided with the outbreak of the Second World War. Nevertheless, F. Mario Boano Scocche per Automobili (aka FMB) continued to produce wooden body bucks and complete 'raw' bodies for coachbuilders scattered around Turin.

Boano enjoyed close family ties with Giacinto Ghia and became co-owner of Carrozzeria Ghia in 1944 following its founder's demise. Even so, Boano continued to operate his own eponymous concern in parallel. Following his well-publicised falling out with Luigi Segre, he departed Ghia and concentrated on transforming his existing business into a rival coachbuilder. He was supported in this endeavour by his son Gian Paolo and future Ghia president Luciano Pollo who had hitherto been financial director of Carrozzeria Ellena. FMB was later liquidated and replaced by two firms: Boano SpA and Boano Lavorazioni Specialia.

Like so many other coachbuilders of the mid-1950s era, Boano's bread and butter came from gussying-up production fodder, Fiat 600 and Lancia Appia elaborazione editions pulling in much-needed lira. Nevertheless, in addition to visual tuning of that nature, Boano's companies produced several celebrated one-offs and small-series runs based on Abarth, Alfa Romeo 1900 and Ferrari platforms. His name is inextricably linked to the Maranello marque because Boano is known for producing a series of 250GT coupés as a subcontractor to Pinin Farina. The irony is that Boano himself was a hugely gifted stylist whose resumé included several famous outlines for which he never received credit. Most notable among these is the Lancia Aurelia B20 GT which features prominently in Pinin Farina's back catalogue.

In 1957, the father-and-son Boano partnership was appointed by Fiat to set up and manage its new Centro Stile department. Boano Lavorazioni Specialia assets passed to Pollo and Ezio Ellena. The sister business Boano SpA existed for another two years but in name only, and then it was liquidated. The workshops at 75 Via Arnaldo da Brescia, Turin began operating under the Ellena Lavorazioni Specialia banner prior to relocating elsewhere in the city as it took on sub-contract work for companies including Carrozzeria Viotti. Ellena folded in 1966.

Felice Mario Boano, meanwhile, retired from Fiat in 1966. He died in 1989.

● Boano (left) was a pivotal figure in the Ghia story, and a brilliant designer in his own right, even if his contributions to cars such as the Lancia Aurelia B20GT, as seen here, went unrecognised *David Cavaliere/The Italian Tribune*, *Giles Chapman Library*.

Chapter Seven
Supersonic: origins of the species

Were it not for its otherworldly looks, this car would have been a mere footnote in motor sport records. The 'Conrero 1900 Coupé Ghia' – one of countless names it was given in print in period – did not cover itself in glory when it competed for the first time on the 1953 Mille Miglia. Fielded and driven by the man who commissioned its construction, Swiss garage owner Robert Fehlmann, car No 453 departed Corso Venezia, Brescia at 4.53pm on 26 April. Ably assisted by navigator D. Duville, the car had pace on its side. However, it had only been completed the evening before the event, so there had been little in the way of shakedown testing. The shapely coupé didn't make it to the finish on the Viale Venezia, having left the course en route and connected with something immovable. The occupants emerged unscathed, but the car was badly damaged.

Despite its off-course excursion, the Conrero Alfa Romeo attracted plenty of attention from the global media. *Road & Track* magazine's report on the event included more photos of the one-off special than any other car …including Giannini Marzotto and Marco Crosara's victorious Ferrari 340MM Vignale. More by happenstance than planning, *The Autocar* magazine managed to photograph the car a few days before the start of the 1000-mile road race. Its 26 June 1953 issue featured a raft of Italian coachbuilt cars, including the Bertone BAT 5 Alfa Romeo styling study together with the similarly powered Conrero. It reported:

> Built by an Italian aircraft engineer, Virgile [sic] Conrero of Turin, for the Swiss driver Robert Fehlmann, the Alfa Conrero sports coupé made its debut on the Mille Miglia, but not unnaturally, as it was completed only a few days beforehand, suffered from teething problems. It is, however, likely to figure in speed events during the present racing season. *Signor* Conrero has previously constructed a number of racing and sports specials. In this latest production, he selected the units he thought most suitable, the Alfa Romeo C, or Sprint-type, 1900 engine, the Lancia Aurelia transmission, rear axle with its independent suspension, and steering, and the Fiat 1400 independent front suspension, mounting them on a tubular chassis.

The article went on to add:

> The engine normally has a power output of 95bhp, but this has been raised to 130bhp at 6400rpm. The modification mainly responsible for the increase is the fitting of four Del' Orto [sic] carburettors, each of which feeds one cylinder. The chassis is formed of two longitudinal steel tubes of large cross-section, united by tubular transverse members welded and brazed together. The centre section of the longitudinals is dropped very low, so that the seating and the overall height of the complete car can also be kept as low as possible.

 Ghia intended building as many as 50 of these 8V-based Supersonics, but the scheme was scuppered by Fiat which was concerned by the lack of customer support in the USA *Michael Bailie*.

Jaguar XK 120 Supersonic

Supersonic: origins of the species

- The 8V was Fiat's halo car in the early-mid-1950s. It remains the only production car in the marque cannon ever to feature a V8 engine, one that was originally conceived for use in a stately limousine *Richard Heseltine collection.*

> Wealthy gentleman drivers were not clamouring for replicas. It was, after all, still unproven as a racing car.

'From the longitudinals, vertical tubes are united by an arched member to form a rigid support for the scuttle. The radiator is mounted very low and inclined slightly backwards. Actually, the header tank is lower than the cylinder head. In front of the lower part of the radiator is an oil-cooler. At the rear, the chassis carries the dry single-plate clutch, four-speed gearbox, and the final drive unit from the Aurelia, which also receives the triangulated suspension arms running from the swinging half-axles. On the very low-hung chassis is mounted a Ghia sports coupé [body] which is particularly low-built as the overall height is only 3ft 11in. The wheelbase is 7ft 11.5in, and the track 4ft 4.75in at the front and 4ft 4in at the rear. The overall length is 13ft 11in and width 6ft, but the adequate ground clearance of 6in has been maintained. The weight in running trim is about 15cwt (1760lb), giving a power to weight ratio of 13.6lb per hp. In its first tests, the Alfa Conrero is said to have attained a top speed of 140mph (225km/h) at an engine speed of 6500rpm.'

Tellingly, there is no mention of Giovanni Savonuzzi in the article, nor any reference to the word 'Supersonic'. That was applied retrospectively. As for the *Autocar* quote 'It is, however, likely to figure in speed events during the present racing season…', the car was rebuilt but with a more conventional open *barchetta* body, before making sporadic appearances in competition to 1955. It then disappeared completely for 30 years. The car was discovered in 1985 and lightly restored by former Conrero mechanic Mario Cavagnero, before being sold to Italian comedian and actor Renato Pozetto. A keen collector of historically important Alfa Romeos, Cavagnero retained the Conrero until 1992. Nothing has been heard of it since.

According to the well respected book *Giulietta da Corsa* by Donald Hughes and Vito Witting da Prato, the Conrero 1900 Coupé Ghia was the first Alfa Romeo-powered car ever tuned by Virgilio Conrero. Other credible sources insist Carlo Abarth built the engine and Conrero merely tweaked it. Whatever the truth, the Mille Miglia accident and the car's subsequent makeover should have marked the end of the story. It was certainly conceived for motor sport use but wealthy gentlemen drivers were not clamouring for replicas. It was, after all, still unproven as a racing car.

However, the Conrero had been briefly displayed on the Ghia stand at the 1953 Turin motor show prior to its departure for Brescia, and its futuristic styling caused a furore. Savonuzzi and Luigi Segre were quickly convinced there was demand for a small-

● The standard 8V (right) with the Rapi outline may not have been exactly pretty but it was distinctive *Giles Chapman Library*. Coachbuilders such as Ghia, Zagato and Vignale rushed to offer alternatives. Brilliant designer Dante Giacosa (below) left an indelible impression on Fiat and the motor industry as a while *Giles Chapman Library.*.

Jaguar XK 120 Supersonic

Supersonic: origins of the species

● Tuning firm Siata worked closely with Fiat on the 8V before creating several iterations of sports cars and GTs based on the car, including this Vignale-bodied offering *Richard Heseltine collection.*

series run of road cars boasting the same outline. Rather than build another lightweight racing car, though, Ghia would now use a proprietary chassis and corresponding running gear, and Alfa Romeo, Maserati and Lancia platforms were considered. Finally, though, the Fiat 8V was chosen as a donor car because it was relatively affordable. This was an inspired move.

The sublime *Otto Vu* was the Turin giant's first – and, to date, only – V8-powered production car. By Fiat standards, 'production' is a relative term, for only 114 were made during a two-year run. Even then, there followed endless permutations and sub-species. In marque terms, the 8V was barely a blip on Fiat's radar. It never made much of an impression on business plans, and must have been expensive to undertake. It certainly didn't return a profit. But that was never really the point of the exercise. In modern day marketing parlance, it was a 'halo product', the reflective glow from which radiated over all the lesser cars Fiat made.

This intriguing *gran turismo* was also a testbed; an exercise in allowing company technicians a little playtime. Fiat first proposed a V8 model in the late 1940s as a flagship saloon – codenamed Tipo 106 – that was serious enough to make it as far as the running prototype stage. However, test hacks failed to impress senior management, and corporate enthusiasm soon melted away. The question then was what to do with the leftover engines that had already been built? The answer was simple: build a GT car.

Enter the *Otto Vu*. Unveiled at the 1952 Geneva motor show, this daring coupé married technical bravado with styling purity in one enticing package. Almost every brand new car carries over at least some DNA from its predecessors, and sports cars generally utilise running gear from prosaic saloons. That was not the case here, though. The 8V was all-new from end-to-end. Catching everyone off guard (there were no leaks or pre-launch publicity prior to the unveiling), this was Fiat brazenly showing off.

Conceived by the brilliant engineering head Dante Giacosa, the powerplant was an over-square, short-stroke unit with 70-deg between banks. This all-alloy jewel featured pushrod valve-gear and a pair of twin-choke Weber carburettors. Good for 105bhp at 5600rpm, and with a relatively modest 108lb ft of torque at 3600rpm, drive went to the rear wheels via a four-speed floor-shift with synchromesh on second, third and top. Of semi-unitary construction – separate steel chassis with tubular side-members, welded to the body – suspension copied the layout of the lowly Fiat 1100, with wishbones and coils at both ends.

Clothing this ensemble was a graceful outline created in-house by Fabio Lucio Rapi and honed in a wind tunnel. With a rear-three-quarter form similar to the Cisitalia 202 Gran Sport's, the frontal treatment was altogether fussier, although the unpleasant art

● Ghia's original rendering for the Supersonic looks stumpy and derivative compared to the finished article *Giles Chapman Library*. Below: all production Supersonics, this Italian-registered 8V included, featured foglights inset in the grille *Motorsport Images*.

Supersonic: origins of the species

● An 8V Supersonic on public display (above), the natural daylight showing the complex interplay of its lines *Motorsport Images*. The only Aston Martin DB2/4-based Supersonic (left) was the last in the series. Some claimed, incorrectly, was bodied in glassfibre *Richard Heseltine collection*

deco grille inflicted on the show car thankfully did not transfer to the production model.

The plaudits soon rolled in. Testing a lightly tuned 8V destined for the Mille Miglia, Gordon Wilkins of The Autocar gushed:

> Handling and road behaviour leave an impression of a thoroughbred with stability, liveliness and immense power. There is general light understeer. The steering is high-geared, and the car goes into a four-wheel drift with all tyres squealing and the throttle wide open in a way which the expert driver will find greatly reassuring.

Predictably, the competition crowd flocked to the *Ottu Vu*. The ready availability of a powerful home-grown engine attracted several small-scale manufacturers, in particular Siata. This Turin minnow used the V8 in its delicious 208S while also assisting Fiat in extracting more power from the three-bearing motor. In time, up to 127bhp was found by virtue of a hotter camshaft and a triple twin-choke carb set-up.

After the first 34 cars had been made, Fiat initiated a restyle. The most obvious modification was an awkward-looking four-headlight arrangement. Predictably, outside coachbuilders were also drawn to the Fiat, with Pinin Farina and Vignale both bodying examples. Even Fiat made a special show car, a glassfibre-bodied coupé that reprised Rapi's standard outline. Then there was Zagato's offering, the Milan coachbuilder crafting its first 8V in 1952. The following year it also presented an open *barchetta* before putting the earlier GT into strictly limited series production. Nonetheless, this being Zagato, no two cars were ever truly alike. Front and hind treatments often diverged; some cars featured the trademark 'double bubble' roof, supposedly on the grounds of adding extra headroom. In 1955, Zagato bought several leftover 8Vs from Fiat and carried on crafting cars into the late 1950s.

They were leftover because the 8V had been quietly axed from Fiat's line-up. This was understandable given that it was almost entirely hand-crafted

● The Aston Martin Supersonic was displayed at the 1956 Turin motor show. Confusion surrounds the identity of its original owner, some sources claiming it was Paris-born American racer Harry Schell *Richard Heseltine collection.*

Supersonic: origins of the species

● Above: the first Fiat 8V-based Supersonic broke cover at the 1953 Paris motor show (above), the outline from the original Conrero car brilliantly transposed on to the new platform *Richard Heseltine collection.*

in-house at a vast cost. Meanwhile, the strictly coachbuilt nature of even the regular catalogue model necessitated a laborious production process totally at odds with the company's ethos of carefully costed mass-production. Similarly, the engine required specialist tooling and could never be mass-manufactured in anything approaching serious volume; at least, not economically anyway. Add in a list price perilously close to three million lira, and a car with a humble Fiat badge had to compete with every elite sports car marque in the business. The 8V, wonderful though it was, was anything but logical.

Ghia took delivery of 14 8Vs in rolling chassis form in 1953 and early 1954. Transposing the Conrero's styling intact on to a new framework was not the work of a moment, but Ghia's artisans persevered. The result was a masterpiece, and became one of the darlings of the November 1953 Paris motor show following its big reveal. *Road & Track* reported on the unveiling in its January 1954 issue, commenting: 'A small 2-litre Fiat V8 chassis with striking two-seater *Berlinetta* body by Ghia gives an impression of a much larger size'.

Wealthy people of taste flocked immediately to buy one of the cars that, somehow, gained the nickname of 'Supersonic'. No-one today knows who coined the name, or why, although the jet-fighter-like contours of the body would certainly have chimed with the public fascination for Mach 1 air flight at the time.

Designer and all-round style maven Howard 'Dutch' Darrin acquired two cars following a visit to the Ghia factory in late 1953. They were displayed at the following year's New York Auto Show, with one sold off his stand to movie star Lana Turner (who subsequently traded it in against a Kaiser-Darrin). The other car was allegedly purchased by millionaire sportsman, Le Mans veteran and car collector Briggs Cunningham, but this remains a source of conjecture among historians. General Motors' head of interior design Henry Lauve (born Henri de Segur Lauve) acquired his Fiat Supersonic at the 1953 Paris motor show.

Serial car builder, Indy 500 entrant and hydrofoil racer Lou Fageol was another customer. He was

- Above and right: well-known classic car dealer and Cobra fan Rod Leach owned the Willment Supersonic hybrid during the 1970s, seen here with Mrs Leach and their young family in 1979 *Rod Leach*.

Jaguar XK 120 Supersonic

Supersonic: origins of the species

● The 7-litre, twin-carb Ford V8 looked out of place in the Willment-modified Supersonic. According to a John Willment Racing Organisation advert from 1967, it produced 420bhp, but other sources claim it was closer to 485bhp *Rod Leach.*

notorious for creating racing cars with engines at either end, including a bizarre Porsche-powered contraption; these mirrored the twin-engined buses made by his family business in Kent, Ohio, USA. 'Lead Foot Louie' bought a Ghia Supersonic for himself and another for his wife, the latter accessorised with prominent tail-fins riveted on top of the existing rear wings, and an incongruous 'continental kit' with an exposed spare wheel. His own Supersonic gained ugly 'bi-plane' bumpers and a Pepco supercharger. With an air of predictability, some Ghia 8Vs received engine transplants, and one of the Fageol cars was soon equipped with a Chevrolet Corvette small-block V8.

Fageol's wasn't the only Ghia Supersonic to receive an American V8, either. A Supersonic body was later dropped on to a Shelby/AC Cobra 427 chassis procured by John Willment. However, this well-known Ford dealer, specials builder and race team patron lost interest in this car even before it was completed, and it has since entered into legend for all the wrong reasons. Apparently, its unruly road behaviour bordered on the lethal.

Whereas Zagato and Vignale sold virtually all their wares close to home, the vast majority of Ghia's Supersonics were bound for the USA. That fact, and the high-profile publicity surrounding their American customers, ultimately hobbled the car's chances. Ghia's original plan was to build up to 50 of its 8Vs but as the *Ottu Vu* itself was never sold in the USA, Fiat was acutely concerned there was no service back up for such cars. If anything mechanical went wrong with them, negative publicity could sully the mainstream models it wanted to market in North America. Ghia enjoyed a close working relationship with Italian-born Detroiter Paul Farrago, and he was mooted as the de facto Supersonic servicing department in the USA, but this was not sufficient to placate Fiat. Ghia relied on Fiat to supply base cars, so it had little choice but to curtail its US ambitions. It could, conceivably, have 'done a Zagato' after Fiat ended production and acquired some surplus 8Vs. But Ghia had already moved on.

● The 1960s-era cubist dashboard was at odds with the torpedo-like 1950s exterior. According to a Classic & Sports Car magazine road test, the cabin was 'workmanlike' and '…neither spacious nor particularly sporting' *Rod Leach.*

Having adapted the Savonuzzi outline from the Conrero Alfa Romeo to a Fiat 8V basis, it set about bottling lightning a second time, only this time using a Jaguar XK 120 chassis. Three left-hand drive Supersonics were built at the behest of the Jaguar importer for France, Royal-Elysées. One of them is the star subject of this book. The conversion was no easy task given the significant dimensional differences between the *Otto Vu* and XK 120 donor cars. There was only a 3in difference in width but the British car was almost 14in longer than the Italian one.

That should have been the end of the story, except that Ghia completed a final variation on the Supersonic theme two years later, this time based on an Aston Martin DB2/2 MkII chassis.

Confusion engulfs the car's construction. Some sources cite racing driver Harry Schell. Paris-born American 'Happy Harry' did undoubtedly use the car, as he drove the Aston to its public unveiling at the April 1956 Turin motor show. In its event report *The Autocar* commented: 'One of the most elegant bodies in the show is the Ghia coupé on an Aston Martin chassis, finished in an exquisite pale green'. In other media coverage there was mention of the car being bodied in glassfibre, patently an error or misunderstanding.

What is certain is the car was sold immediately after the show closed. It was acquired by Richard C. Cowell, a friend of Aston Martin proprietor David Brown. Cowell, a record-breaking water-skiing champion and all-round jetsetter, paid $15,000 for the car and, according to the logbook, he and not Schell was the first registered owner. In 1958, Cowell married American socialite and heiress Gail Whitney Vanderbilt and gave the Supersonic to his new bride. They divorced a year later and the car was subsequently acquired by dealer and gentleman racing driver Bob Grossman. It sold for $2,310,000 at RM Sothebys' 2013 Art of the Automobile sale in New York. That wouldn't be the last eye-watering figure achieved by a Supersonic.

Chapter Eight
Jaguar XK 120: foundations of a legend

The Jaguar XK 120 - or XK Open Two-Seater Super Sports as it was initially dubbed – caused a seismic impact when it was unveiled at the 1948 Earl's Court Motor Show. The world's media was sent into a tailspin, even though marque founder William Lyons didn't foresee the car having a future. The car was a rolling laboratory, a testbed for an all-new engine. It wasn't intended for volume production but demand for Jaguar's exciting new sports car was such that the coachbuilt, limited edition approach soon became untenable. After the first 239 cars had been completed, hand-beaten aluminium bodies (over laminated ash wood frames) made way for pressed-steel lookalikes, with aluminium doors, bonnet and boot lid skins. Even then, demand outstripped supply by a significant margin.

All of which is easily explicable: the XK 120 was that rarest of things – a bargain. Priced at £998, there was nothing approaching its specification, or anywhere near as fast, for the money. Powered by Jaguar's new twin-cam, six-cylinder engine, the 120 part of the nomenclature signified the ultimate velocity in miles-per-hour (the X was for experimental, the K for the sequence of engine design designation). For the same outlay, you could have bought a Lea-Francis 14hp Sports which, for all its virtues, somehow didn't have quite the same allure. The XK 120 was swish, sexy and elegant, all at the same time.

Arbiters of beauty routinely cite the XK 120 as one of the greatest feats of automotive artistry ever perpetrated, but the outline wasn't entirely original. Certainly, Lyons had a great eye for line and proportion, but he was not a designer in the strictest sense. By most accounts, he couldn't draw particularly well and, like most of the greats, worked principally by instinct. Lyons' real talent was his ability to appropriate the bits he liked from other cars and then improve upon them, and this magpie approach was obvious with the XK 120. The car's silhouette bore more than a passing resemblance to the Carrozzeria Touring-bodied BMW 328s that won the distance-shortened Mille Miglia in 1940. Not a single line was directly cribbed; they were just processed, polished and perfected. Legend has it that Lyons rustled up the XK 120 outline in just a fortnight. There were no committee meetings or customer clinics for him.

Aside from its fabulous styling, the XK 120 also excelled in motor sport, and not just in Britain. Lyons' creations had enjoyed competition success before, in particular Jack Harrop's victories with the SS100 on the 1937 and '38 RAC Rally of Great Britain events. But the breadth of the XK 120's achievement far outstripped even that of pure-bred racing exotica.

Predictably, a number of aluminium-bodied examples found their way into motor sport, the likes of Leslie Johnson, Peter Walker, Clemente Biondetti, Tommy Wisdom and Ian Appleyard being among early adopters. The last named – Lyons' son-in-law - in particular heaped glory on the brand, achieving an unprecedented Gold Cup hat-trick on the International Alpine Rally from 1950-52, along with RAC Rally of Great Britain and Tulip Rally honours in 1951, always driving his well-known car with registration number NUB 120, the sixth of six cars works-prepared for competition purposes. Stirling Moss' triumph on the 1950 Tourist Trophy at a

● The Jaguar 'XK Open Two Seater Super Sports' was – and remains – a styling masterpiece. It established the Jaguar template of striking looks and high performance at an affordable price; little wonder demand soon outstripped supply *Jaguar/Newspress.*

Jaguar XK 120 Supersonic

Jaguar XK 120: foundations of a legend

● The XK 120 established Jaguar on the world stage. It was both devilishly attractive and a prolific winner in races and rallies. It also helped cement marque instigator William Lyons (right) as a global influencer *both Jaguar/Newspress.*

waterlogged Dundrod – on the eve of his 21st birthday – further cemented the XK 120's stellar status on the world stage.

This was underlined once the fixed-head coupé variant broke cover in the spring of 1951, followed by the luxurious drophead coupé two years later. While the world's supply of superlatives had seemingly been exhausted with the roadster, the closed edition had a different, more urbane agenda. Here was a car intended for high-speed travel over epic distances in saloon-like comfort, rather than for energetic back-road jollies. This point was hammered home in August 1952 when Leslie Johnson – who had finished fifth on the 1950 Mille Miglia in his privateer XK 120 roadster – embarked on a seemingly impossible mission. He had already broken two significant long-distance records, averaging 107.46mph for 24 hours in 1950 along with Stirling Moss, and then 131.83mph over one hour a year later; now he set about averaging 100mph for an entire week.

Armed with the second right-hand drive XK 120 coupé ever made, one with few modifications save for extra spotlights, a supplementary fuel tank and a two-way radio, his was an audacious bid. The sometime Grand Prix driver took to the banking at L'autodrome de Linas-Monthléry near Paris, with Moss, Jack

● Both attractive and reasonably practical, the XK 120 was also by far the fastest production car of the day in its price range *Jaguar/Newspress*. The fixed-head coupé model (below) followed in early 1951, with the likes of racers Stirling Moss and Mike Hawthorn among its cheerleaders *Giles Chapman Library*.

Jaguar XK 120 Supersonic

Jaguar XK 120: foundations of a legend

● Right: Ron 'Soapy' Sutton guided this Jaguar XK 120 to just over 132mph in May 1949 to claim a new speed record for a production car *Jaguar/Newspress*. Below: this XK 140 FHC was privately entered in the 1956 Le Mans 24 Hours and driven by Roger Walshaw and Peter Bolton. It was running in eleventh place when it was erroneously black-flagged for a refuelling infringement during the 21st hour *Jaguar/Newspress*.

● Above: The XK series was rounded out by the XK 150 which was on sale from 1957 to 1961 *Jaguar/Newspress*. Below: From coachbuilder to mass manufacturer, Sir William Lyons is seen here with his wife, dogs, the SS 1 and an XJ12 in 1972 *Jaguar/Newspress*.

Fairman and Bert Handley also on the roster, and together they lapped the circuit more than 12,000 times over the course of 168 hours and achieved Johnson's goal by averaging 100.31mph. The all-British team left France with five new class records and also snared four world records for good measure.

Given the XK 120's ability to marry high speed with civilised comfort, it is little wonder that the likes of Moss, Jack Sears and Ian Stewart used coupés as their road transport. They were among rarefied company, because only 152 examples stayed in the UK while the 2500 or so others were earmarked for export. With the arrival of the XK 140 in 1954, the bloodline continued. The even bigger-boned XK 150 appeared in 1957. It was only ousted by the arrival of the E-type in 1961.

William Lyons' capacity for understanding mass appeal without trading integrity was without equal; the XK 120 had substance below the surface flash. He didn't do formulaic box-tickers. His willingness to use motor sport as a platform also served to advance the brand on the global stage. It's hard to escape the gravitational pull of the XK 120's trackside legacy because, without it, there would have been no C- or D-type.

Jaguar XK 120 Supersonic

Chapter Nine
Latin interpretations: the other Jaguar XKs with Italian bodywork

Jaguars have always been cars whose styling has overshadowed rivals, and it is a brave designer who thought he could do better. However, that didn't stop several Italian *carrozzerie* from trying to improve upon the XK 120, XK 140 and XK 150 to tempt 1950s buyers.

Bertone Jaguar XK 150

Carrozzeria Bertone's Franco Scaglione was one of the great free thinkers of automotive styling; part sculptor, part aerodynamicist, and something also of a boundary-pushing visionary. His take on the XK 150 was, nonetheless, restrained by his own avant-garde standards. There was no reaching for the stars here. Nonetheless, it was well-received when displayed at the 1957 Turin motor show. So much so, in fact, it led to two more cars being completed, purportedly to hotter XK 150 S-spec. One was UK registered, but the identity of who commissioned it remains unrecorded. At the time of writing, the sole remaining car resides in the Netherlands. Bertone, a coachbuilder and styling house of great acclaim, later reworked a raft of Jaguars to create show cars based on the E-type, S-type 3.8-litre, and XJ-S platforms.

Zagato Jaguar XK 140/XK 150

Milanese coachbuilder Zagato created several landmark designs during the 1950s, but its individual take on the XK series remains a subject of confusion and conjecture. It was once widely held that three cars were made, but credible historians have adopted the view that only two were so bodied.

The first example was built on an XK 140 platform at the behest of Italian playing card manufacturer Guido Modiano, who had crashed the donor car. The finished article was displayed at the 1957 Paris motor show. A second example, based on XK 150 running gear, was then built for a Swiss Jaguar dealer and exhibited at the 1958 Geneva show. The alleged third car, if it ever existed, would have been based on an

● Left, above and below: Zagato's take on the XK theme was restrained by the design house's outré standards *Richard Heseltine collection /Alessandro Sannia*. Right: Bertone's offering from the pen of Franco Scaglione was similarly muted *Richard Heseltine collection*. Only tiny numbers of both designs reached the road.

Jaguar XK 120 Supersonic

Latin interpretations: the other Jaguar XKs with Italian bodywork

XK 150 SE. A lavish brochure was produced in period, which suggests a production run may have been mooted, but Zagato's own records are too scant to include any evidence of such a scheme.

Stabilimenti Farina Jaguar XK 120

Stabilimenti Farina rebodied three Jaguars in 1951-52 for the Belgian marque concessionaire Joska Bourgeois. Her Anglo-Belgian Motor Company commissioned the construction of two cars based on the MkVII saloon and also an XK 120 which received a sober-looking coupé outline penned by Franco Martinengo (although it is widely held that it was derived from a previous design by Giovanni Michelotti). The dark red 'Flying Jaguar Coupé' was the very last car fashioned by the 47-year-old firm founded by Giovanni Carlo Farina. The doors of the premises were shuttered in 1953, Martinengo having already departed for Pinin Farina.

Michelotti Jaguar XK 140 SE

Giovanni Michelotti reputedly bodied three Jaguar XKs, none of which were an improvement over the Lyons original. The best known of these was a car built in 1957 and based on a two-year coupé which had been badly damaged in an accident. It was later acquired by Roland Urban, one-time owner of the subject of this book, who claimed it had once been the property of silver screen goddess Bridget Bardot. She denied having ever acquired a Jaguar of any kind (and did so via her lawyer). It was later sold to a Belgian enthusiast who didn't complete the restoration, and instead put the car into storage. It sold for €365,000 at a 2018 Bonhams auction, which was more than ten times the pre-sale estimate. Michelotti later re-bodied a Jaguar D-type which was beautifully proportioned, although it was subsequently robbed of its racing car componentry.

Ghia Jaguar

Not content with creating a trio of Jaguar Supersonics, Ghia followed through by building a small series of unlovely Jaguar XK 140-based coupés. Three are believed to have been made, the last in 1955

● Stablimenti Farina bodied Jaguars (above) on behalf of the Belgian concessionaire, and none was prettier than the donor cars *Barry Welch*. Michelotti's offering (left) once shared a private museum with the subject of this book.

to the order of Jaguar's Lebanon distributor Robert M. Trad. Another was once retained by Mexican-born Hollywood heartthrob Ricardo Montalbán, while the Aga Khan is known to have owned one example. Confusion surrounds the identity of who styled the Ghia XKs, with prolific pen-for-hire Giovanni Michelotti a likely candidate. Accordingly, some sources may have unwittingly conflated the histories of the Ghia cars with those created by Michelotti under his own name. The once closely-associated Swiss Ghia-Aigle concern also produced a restyled XK 120 with an ugly fixed-head roof which was used in rallies, plus a brace of XK 150-based, Maserati-esque GTs penned for the coachbuilder by Pietro Frua.

Motto Jaguar XK 150

Rocco Motto was never much of a name in the global design world, but his Turin bodyshop was once a hive of activity. From 1932, he worked extensively as a subcontractor to Carrozzeria Ghia and Stabilimenti Farina, among others. In the immediate post-war years, he made bodies for Cisitalia and also worked hand in glove with Giovanni Savonuzzi on the SVA single-seater. Unusually for the period, he often bodied foreign cars, from Delahaye to Packard via Cadillac and Renault. In 1957, he and his brothers Ernesto and Clemente bodied an XK 150 with a coupé outline which resembled a contemporary Bristol. A

● Above: After constructing three Jaguar-based Supersonics, Ghia followed through with a small run of re-clothed XK 140s, although Giovanni Michelotti may have styled them as a freelancer *Motorsport Images/Richard Heseltine Collection/Alessandro Sannia*. Right: Motto's less than lovely take on the XK 150 resembled a contemporary Bristol *Alessandro Sannia*.

Latin interpretations: the other Jaguar XKs with Italian bodywork

● Pinin Farina shaped XK 120 (above) may have been built for influential Austrian-American motor mogul Max Hoffman *Newspress*. Industrial designer Raymond Loewy's Boano-bodied XK 140 (right) attracted controversy in period *Richard Heseltine collection*

second car was constructed near concurrently but with a different grille.

Pinin Farina Jaguar XK 120 SE

The precise backstory of this one-off *gran turismo* is lost to history but its current owner claims the car was built new for the influential Viennese-born New York motor mogul Max Hoffman. In 1954, he purportedly instructed Pinin Farina to construct 'Project PF471', the finished article breaking cover at the March 1955 Geneva motor show. The car was pictured in *Road & Track* magazine in June 1955 but later disappeared from public life, only for a German collector to discover and purchase it in derelict condition in 1978. He never got around to restoring the car but retained it until 2015 when ownership passed to a British enthusiast. More than 6000 hours were then sunk into restoring it, since when it's been garlanded at many of the world's premier *concours d'elegance* events.

Boano (Loewy) XK 140

The brainchild of legendary industrial designer Raymond Loewy, this inelegant coupé was rumoured to have been rooted in a stillborn Ferrari project which only progressed as far as the quarter-scale model stage. An XK 140 platform was substituted, with the outline being tweaked to accommodate the shorter wheelbase and loftier engine of the

Coachbuilt Cars

● Serafino Allemano's fame as a coachbuilder may have been fleeting but he was responsible for creating many beautiful cars. His reworked XKs were perhaps derivative of his Maserati outlines, one of the two made being styled by prolific pen-for-hire Giovanni Michelotti *Alessandro Sannia.*

new donor car. The unconventional styling has retrospectively been attributed to Loewy's cohorts John Cuccio (who went on to become head of Ford's Mercury styling department) and sometime General Motors man Orval Selders. There is every reason to believe there were significant deviances between their design work and what was created in three dimensions by subcontractor Carrozzeria Boano in 1955. The end result was displayed at that year's Paris motor show minus bumpers. The car was shipped to New York in 1956 and a year later ownership passed to boxing champion Archie Moore, for £25,000. The car was destroyed in a fire at George Barris' custom car emporium in California in December 1957. Loewy was not done with stamping his mark on Jaguars, though, as he later restyled – some might say ruined – an E-type.

Allemano XK 140/XK 150

Serafino Allemano's small carrozzeria is perhaps best known for bodying assorted Abarths (both front- and rear-engined) plus 22 Maserati 5000GTs. His take on the XK140 from 1956 borrowed heavily from earlier outlines produced for Fiat 1100TC and Maserati A6G 54 running gear. Resplendent in white with a blue roof, the pleasing styling was otherwise spoiled slightly by a bank of air vents sunk into the front wings, and some clumsy detailing. A second car, based on an XK 150 to a design by the ubiquitous Giovanni Michelotti, attempted to integrate a typical Jaguar grille in a silhouette that embraced most of the prevalent styling fads of the day, including a panoramic roof and vestigial tail fins. It was built for a Swiss customer but the car vanished without trace many years ago, along with its sister.

Chapter Ten
Jaguar XK120 Ghia Supersonic: the life and times of chassis 679768

Few Supersonics – if any – have attracted more attention in the specialist press and wider media than chassis number 679768. In period, it appeared at some of the more prestigious European motor shows in addition to racking up generous column inches in magazines the world over. The Jaguar was also a winner at high-end *concours* events when it was a brand new car, and it is still being cited more than half-a-century later as a blue chip and copper-bottomed classic. Design pseuds amateur and professional routinely pour forth the most purple of prose in describing the car and why it matters. Nevertheless, an aura of mystery surrounds its origins. Much of this centres on the man who acquired it. It has been all too difficult to differentiate between whose truth is the most plausible.

This was the first XK 120 clothed by Carrozzeria Ghia. It was one of a pair ordered by the French Jaguar concessionaire Royal-Elysées of 80 Rue du Longchamps, Paris. One of them – this one – was finished in red, and the other blue. Chassis 679768 began life as an XK 120 fixed-head coupé and differed significantly in detail to its sibling. The blue version featured an engine-turned dashboard and matching transmission tunnel rather than a painted duotone aluminium arrangement as here. The red car had fewer bars in its chip-cutter grille, and was constructed without a bonnet bulge and air vents sunk into its flanks.

The most important difference between the two cars, however, was beneath the skin. While the blue car was mechanically standard, the subject of this book received a significant hike in horsepower thanks to a Conrero-tuned engine. The classic Jaguar straight-six was equipped with triple twin-choke Weber carburettors on a Conrero forged-alloy manifold in place of the standard Jaguar's SU carburettors. This, according to some articles of the period, meant it generated as much as 220bhp. That represented a power hike of 60bhp over a regular production XK unit.

Both Jaguar Supersonics were sold new to a *Monsieur* Malpelli of 287 Rue Vendome, Lyon. Labelled a 'colourful Lyons rag-trader' by *Classic & Sports Car* magazine, the mysterious Malpelli manufactured ladies' support hosiery among other garments for the fashion-conscious. The same title hypothesised that he may have acquired the Supersonics for publicity reasons. Jon Pressnell wrote:

> If you wanted to create a stir in the French fashion world of the '50s – even then criss-crossing the automobile world at ultra-chic *concours d'elegance* events – there was one sure-fire way to do it: publicise your latest clothing in the company of a breathtakingly futuristic special-bodied Jaguar, tailor-made for you by a trendsetting Italian design house. But why stop there? Why not splash out on a matching pair of *haute couture* Italian-suited Jaguars?

This Ghia flight of fantasy had the desired effect. In 1954, this red car was displayed at the London motor show. It also wowed crowds at major *concours* events at Cannes and Montreux. At the former, it was displayed

- The Supersonic's shapely rump is dominated by afterburner-style tail-lights. Each lamp comprises 32 specially fabricated components. Two of the four exhaust outlets are dummies
Darin Schnabel.

Jaguar XK120 Ghia Supersonic: the life and times of chassis 679768

- The Jaguar Supersonic's front-end is dominated by the simple but generously proportioned oval air intake and tubular quarter-bumpers. Note the Marchal lamps inset in the grille *Darin Schnabel.*

alongside the other Malpelli Jaguar plus the third of the XK 120 Supersonic trio built, which had been sold to a wealthy Swiss buyer. This marked the only known occasion when all three Ghia Jaguars were seen together publicly. The other Supersonic, resplendent in white, took home the *Grand Prix d'Honneur* prize. Malpelli's beautifully-attired fashion models in attendance were every bit as eye-catching as the gleaming jet age super-coupés they posed beside. That said, one of these ladies may have been responsible for inflicting migraines on historians in years to come.

The red car was registered 69 BJ 75, but appeared at Cannes bearing the legend '66 BJ 75', which served to muddy the waters decades later as attempts were made to disentangle the car's narrative. It transpired that one of the models took exception to the original registration number, in particular the *soixante-neuf* part. She felt it perhaps sent out the wrong message about her sexual proclivities, and Malpelli acquiesced to her request – maybe demand – that it be changed.

A fake plate was substituted.

This anecdote has the whiff of the apocryphal about it, but so much about the enigmatic M Malpelli defied belief. He soon fell behind in paying for the Supersonics and did a moonlight flit to Africa, leaving a trail of debt in his wake. Chassis 679768 was subsequently reclaimed by Royal-Elysées, but not before photos of it appeared in the 1955 edition of the annual *Automobile Year*. It was also illustrated in the 1955 book *Sports Cars*, by John Wheelock Freeman, the author using the Ghia Jaguar Supersonic as an exemplar of the stylist's art. He argued, not always coherently:

' The final question arises as to why it matters whether a car is designed with awareness of its meaning as well of its function. To most who love cars, it frankly does not matter; the car is a tangible object and it appeals primarily to materialists. Their sensitivities are chiefly mechanical and they accept, reject or ignore

the body design as it is without knowing what it is, much as a sculptor would accept, reject or ignore the engine and transmission. At best, they apply external stylistic criteria to details. But since the design is admittedly not of integral importance to them, surely, they will not object if it is well done. So, the possibility of theirs to cooperate is not an obstacle; they are simply, by their own choice, not involved.

Royal-Elysées placed the car in storage in the mid-1950s. It was moved repeatedly prior to 1969 when it was sold as part of a general clear-out after proprietor Charles Delecroix ceased importing Jaguars. The dealership's former workshop manager Michel Cognet recalled in *Classic & Sports Car* in 2000: 'The car had been abandoned, and nobody was interested in that sort of vehicle at the time'.

Malpelli, meanwhile, returned from his decades-long, self-imposed exile and became a litigation-obsessed fantasist, firing off lawsuits to recover his lost fortune. According to the same *Classic & Sports Car* article, he would: '…wave around letters from Edward and Robert Kennedy sent in reply to his "this-is-how-to-put-the-world-to-rights" missives to the American politicians'. Now of greatly reduced circumstances, he lived in a candlelit apartment, his electricity having been cut off, with only ever-growing mounds of legal documents to keep him company. He did, however, manage to sell the blue sister car to a determined fan of the Supersonic – perfectly legally, as it transpired, despite not knowing of its whereabouts. It was later discovered following many years of judicious gumshoeing on the enthusiast's part. The location of the white 'Swiss' Supersonic, by contrast, remains unknown at the time of writing.

The second recorded owner of chassis 679768 was Dr Philippe Renault, who co-founded the French Jaguar Drivers' Club in 1968 and later acted as its chairman. The forensic dentist was a perfect custodian, having owned and raced a variety of significant road and pure competition Jaguars in

● Despite so many seemingly disparate styling elements, the Jaguar Supersonic's styling is cohesive and beautiful with it. The roofline is redolent of some of the Ghia-Chrysler dream cars *Darin Schnabel.*

Jaguar XK120 Ghia Supersonic: the life and times of chassis 679768

Tuned XK 120 engine featured Conrero big-valve cylinder head and triple racing Weber carburettors on a specially made, forged-alloy manifold. It produces near C-type levels of power, which ensures the performance lives up to the car's racy looks *Darin Schnabel.*

historic motor sport including a C-type and an XK SS in addition to several Jaguar-powered Listers. These included a 'Knobbly' variant which he claimed was supplied new to Carroll Shelby, and the one-off, ex-Peter Sargent/Peter Lumsden Le Mans car with bodywork shaped by legendary designer-aerodynamist Frank Costin, which he co-owned with historic racing stalwart John Harper.

Dr Renault's world-class collection of Jaguars and Jaguar-powered sports-racers was displayed in a private museum near Le Mans, but his custody of the Supersonic proved fleeting. In the early 1970s, ownership passed to his friend, Jaguar Drivers' Club president and serial collector Roland Urban. The son of a Russian-born Hungarian immigrant, this remarkable Frenchman was variously a racing driver, author and historian but was better known as a stuntman. He was in his early twenties when he was hired as a parachutist during production of the 1962 Second World War cinematic epic *The Longest Day*, and putting himself in harm's way soon became his profession. Urban used his paycheque from the production to buy a Renault Dauphine that he entered in a rally and promptly rolled into a ball. Nevertheless, the die was cast. From there, he went on to work with director John Frankenheimer on *The Train* among several Hollywood-financed war films, in addition to other independent productions and television dramas. Such was his standing within the film community that he doubled for big screen legend Al Pacino in the Formula One-rooted *Bobby Deerfield*, performing the same on-set duty for superstars Robert de Niro, Cary Grant, Burt Lancaster, Peter O'Toole and Jean-Paul Belmondo. By his own reckoning, he performed stunts in around 300 productions, and also took on occasional acting roles and second unit directing duties.

Urban's demanding day job ensured he was often away on location, sometimes in foreign countries and for long periods, so his passion for cars and motor sport remained a hobby. Nevertheless, he participated in more than 400 races, rallies and

- Trimmed in beige leather, the beautifully detailed interior stretches to a duotone dashboard, and lovely knurled aluminium knobs to control the headlamps. The winged Ghia badge appears on the door cappings *Darin Schnabel.*

hillclimbs spanning five decades. He carried on competing into his early seventies. The vast majority of his outings were in Jaguars or Jaguar-powered sports-racing cars, and inevitably this passion for the Coventry marque spilled over into collecting. Urban had a particular fascination for one-off coachbuilt cars, or variants built in small numbers. In addition to the Supersonic, his private museum near Montlhéry was home to the unique Michelotti-bodied D-type and a Farina-bodied MkVII among several other prototypes and showstoppers. He wrote of his cars in his book *Les Metamorphoses du Jaguar*, published in 1993, although the text largely comprised extended captions.

During his ownership, chassis 679768 remained in largely original condition and was used only sparingly. In 1994, custodianship of the Supersonic passed to another French Jaguar Drivers' Club luminary, Jean-Claude Ferchaud. *Rétroviseur* magazine wrote of him in 2005:

> Whether at the Rétromobile show, or competing on a rally or *concours d'elegance*, not forgetting the route of the Tour de France Automobile, you will invariably come across Jean-Claude. He is a collector for several decades, and has travelled all the roads of Europe aboard countless old cars of all types. At present, Jean-Claude's collection consists entirely of Jaguars. All are preserved in an exceptional state, and he cares more about quality rather than quantity.

Ferchaud's passion for cars and motor sport was encouraged and fed by sometime Grand Prix driver and Le Mans regular Guy Mairesse. As a boy in the 1950s, Ferchaud went to school with his son Michel and they soon became firm friends. They attended race meetings together and devoured copies of motor sport magazines. In 1951, the youngsters listened to the radio broadcast of the *19èmes Grand Prix d'Endurance les 24 Heures du Mans*, in which race Guy

Jaguar XK120 Ghia Supersonic: the life and times of chassis 679768

Precious metal
Stripped bare to future-proof

• During the writing of this book, the one and only Conrero-tuned Jaguar Supersonic underwent a nut-and-bolt restoration at the workshops of Strada e Corsa in the Netherlands. The firm has restored several other Fiat 8V-based Supersonics, in addition to other low-volume Italian exotica.

Jaguar XK 120 Supersonic

Jaguar XK120 Ghia Supersonic: the life and times of chassis 679768

- The Strada e Corsa team disassembled the Supersonic before setting about returning it to era-correct perfection. The current owner aims to display the car at the world's most significant *concours d'elegance* events.

Jaguar XK120 Ghia Supersonic: the life and times of chassis 679768

Mairesse finished second overall alongside Pierre Meyrat aboard their Talbot-Lago T26 GS Biplace. It was a remarkable achievement for two gentleman drivers against major works opposition. For Ferchaud, however, it was the winning car that really captured his youthful imagination. He became enraptured by descriptions over the radio of the Peter Walker/Peter Whitehead Jaguar C-type; even more so after he saw a photo of the Le Mans victor in a newspaper a day later. As someone whose first passion was aviation, he wrote in later years that he fell in love with the shape of the C-type because its streamlined outline reminded him of aircraft fuselages. "I wanted to be a pilot or do something in aviation," he told *Rétroviseur*.

' My parents, by contrast, absolutely did not want to hear about such a vocation so I became an engineer. In my professional life, I built factories all over the world. A lot of them were power generation plants. I did those all over Saudi Arabia; large-scale projects where I was also the project manager. I had done enough to live well, and loved being able to share such prosperity with my family, but I never forgot my first passion – aviation. When I was young, I collected everything I could related to aircraft and aircraft design. I then moved on to making models. As soon as I had the opportunity, I learned to fly and earned myself a pilot's licence.

This came in useful while Ferchaud was working aboard, but then it dawned on him that some of the aeroplanes he flew in more remote climes were not well maintained. "I was in no position to buy and run my own aircraft so I decided I should probably stop". Ferchaud found a suitable substitute in classic cars.

' I had a Ferrari 512BB, and raced a Porsche 2.7 Carrera RS, and then dabbled with all sorts of stuff including an NSU Sport Prinz and a Panhard 24CT. My interest in the 1980s was pretty eclectic, and at one point I had a Porsche 356 convertible by the coachbuilder Drauz, and also a superb OSCA MT4 *barquette* in which I did the first of the retrospective Tour Auto events [in 1992]. However, by then I had remembered an automobile brand from my childhood and Jaguars became my point of focus.

My first was a 1958 3.8-litre XK 150 drophead which is still present in my collection. There is a good chance that it will stay here for a long time. I also had an XJ6 saloon and an E-type coupé in which I toured Europe. It was an exceptional car which only gave me pleasure. Then one day, I learned that Roland [Urban] was determined to sell his famous Jaguar specials, but not the Supersonic. My preference was for the Ghia car, but then there was a real miracle. Eventually, he agreed to sell the Supersonic to me and it was a privilege to be able to acquire the car. At that time, the Supersonic had signs of patina and was partially dismantled. It had done only 9400 kilometres from new. The engine was running

● To infinity and beyond! The Supersonic's 'jet age' styling is perhaps most striking when viewed from above. It's hard to believe this design was chopped and changed from the original Conrero-Alfa racer so effectively
Darin Schnabel.

100   Coachbuilt Cars

very poorly but I was able to make it run. Being an engineer gave me some fluency in these things, but also I did an internship at a local garage when I was 13 years old. My father had been keen for me to learn a trade and I worked for one of his mechanic friends during the holidays. I received an excellent schooling.

Nevertheless, despite his keen grasp of all things mechanical, he left it to the professionals once he'd decided to renovate the car. Jaguar specialist Atelier Sontrop of Pourrain was chosen to perform the makeover. The bodyshell was taken back to bare metal ahead of minor surgery and a repaint in the original metallic 'ruby red' hue. The car also received a complete mechanical overhaul, the Conrero engine being rebuilt but retaining its original cylinder head and carburettors. The interior was similarly rejuvenated rather than replaced, the Ghia-trimmed leather upholstery being kept in situ. Ferchaud, it would appear, was interested in preservation rather than restoration wherever possible. According to receipts from 1995, the entire renovation cost Ferchaud 240,000 francs, although a hand-written note from April 2007 found among the car's history file suggests it was actually closer to 380,000 *francs* when post-restoration servicing and other work is taken into account.

Ferchaud fielded the newly fettled Supersonic in the prestigious Louis Vuitton Concours d'Elegance at Bagatelle. Paris in September 1996. It emerged victorious, claiming class and overall honours. The car also appeared at Rétromobile in Paris in February 1999 where it formed part of a special display celebrating coachbuilt Jaguars.

This was no trailer queen, though, with Ferchaud adding some 4000km to the end of the decade. He wasn't above letting others share his good fortune, either. No Ghia Jaguar was ever independently tested in period, and ex-pat Briton Jon Pressnell was arguably the first journalist ever to drive one when he evaluated Ferchaud's car in May 2000. The *Classic & Sports Car* contributor got behind the wheel. He reported:

● The most beautiful coachbuilt GT of the 1950s? That's debatable, but the Jaguar Supersonic is a sure-fire candidate. Unlike so many other flights of fantasy from the period, this one is also usable in the real world
Darin Schnabel.

‷ Step into the cockpit and you soon realise the Jaguar Supersonic is no impractical dream car, even before you've fired up the tuned XK straight-six. Trimmed in beige leather, it's a spacious Grand Touring environment, uncompromisingly a two-seater but with a large luggage platform behind the seats – a necessity, given that the boot is full of fuel tank and spare wheel. From the handbag-like pouch pockets to the duotone dashboard and the pioneering leather-strap seatbelts, the detailing is exquisite. Take in the slenderly long door armrests, the chrome footrest for the passenger, the lovely knurled aluminium knob for the headlamps, and the winged Ghia badge on the door cappings and, as you install yourself behind the low-set Nardi wheel on its ruby-red painted column, you can agree with Jean-Claude's assessment of Ghia's approach to interior décor – "make it simple, but make it elegant".

Jaguar XK 120 Supersonic

Jaguar XK120 Ghia Supersonic: the life and times of chassis 679768

The doors shut with a feeling of solidarity and precision, and the panel shut lines are nicely even without having had to be seriously fettled during restoration. But Ghia didn't hang around when building the Supersonics, and so wing edges are simply folded rather than wired, just as the bonnet has straightforward folded edges and a plain square-tubed reinforcing member, and the boot panel is single-skinned. On the other hand, the handmade "F" on the boot is a delight, and the tail lamps each comprise 32 beautifully fabricated components, the whole ingeniously clamping to the bodywork… Making a more attractive Jaguar than a Lyons original is a challenge that over the years most stylists have muffed. The exuberant Supersonics have to count as the few exceptions…

Driving holds few surprises, other than the extraordinary tractability of what is close to a C-type tune engine running on racing Webers with fixed idle jets. Trickling through Paris traffic at 1500rpm in third, the car picks up speed without hesitation, and also keeps its cool – due, one suspects, to the generous grille opening – during prolonged stop-start crawling. With the road clear, and the Jaguar beefily hitting a more appropriate and hot oil-tinged 70mph, the deep-throated and robustly barking note of an insistently-used, tuned XK is as you would expect, as too is the need to go gently with the long-throw Moss gearbox,

● Jaguar's onetime mantra of 'Grace, Space and Pace' is applicable to the XK 120 Supersonic. It rendered the donor car positively pedestrian by comparison, the roofline here appearing improbably low
Darin Schnabel.

Jaguar XK120 Ghia Supersonic: the life and times of chassis 679768

which even Jean-Claude manages to graunch into second. On standard XK 120 drums, the brakes have a firm short-travel action and cause no concern. It's only with hard or more prolonged use that they soften up, says Jean-Claude. With the XK's conventional leaf-sprung rear left unmodified, the suspension is somewhat abrupt at the back, where the car is apparently lighter than a regular XK 120, but otherwise it sends out the same undemanding messages'.

Ferchaud continued to use the Supersonic, often participating in French Jaguar Drivers' Club events, until May 2007 when it went under the hammer at Bonhams' Les Grandes Marques A Monaco sale. It sold for €753,000 (£677,044) including premium. Ownership passed to Swiss classic car dealer and proprietor of Graber Sportgarage AG, Christian Traber. He used the car sparingly, not wanting to add too many miles to the odometer that read at just shy of 22,000km (13,640 miles). Nevertheless, Traber allowed Belgian reporter Dirk de Jager to drive the car for an article for American car magazine *Autoweek* in 2012 (which was later syndicated to *Routeclassiche* and *Der Automobiel Klassieker Magazine*). Remarkably, he was able to pitch it against a Fiat 8V Supersonic in the same colour for the article. Both belonged to Traber. De Jager wrote:

❝ Getting behind the wheel of the XK 120 is a bit of an experience for anyone more than six feet

Jaguar XK120 Ghia Supersonic: the life and times of chassis 679768

● Boot space is at a premium thanks to the spare Borrani wire wheel and the fuel tank. The filler cap is impractically placed but its concealed location doesn't interrupt the styling flow
Darin Schnabel.

tall as there isn't as much room behind the big wooden steering wheel as you might expect. A comfortable seating position is difficult to find because of the cramped conditions. We have to push the throttle down from the bottom of the pedal, but luckily it's mounted from above, making it a bit easier. When first gear is engaged, the lever is struck against my knee and gives no room at all to move in order to operate the pedals. Even headspace is cramped, but with a car that looks this good, you just grin and bear it. Once we are moving, it comes as a bit of a surprise how heavy the steering is, but again, you adapt to it quickly, and after a few minutes I didn't really notice it any more. It is only later in the evening that you feel it in your arm muscles that you had a bit of a workout in the car. Yet cruising around the Swiss Alps following the similarly designed 8V is just an out of this world experience. Savonuzzi's design attracts attention from nearly everyone who sees it, and seeing two such wildly designed cars driving together is truly something special.

Correspondence dated January 2016 confirms that the car was rarely seen publicly during Traber's ownership. He wrote:

> The only *concours* I attended with the car was the 2014 Chantilly Arts and Concours d'Elegance. This was the first edition of this new concours, which is probably one of the best in the world (close behind Pebble Beach). I also did the rally the day before the *concours*… I finished second with the car but got the special prize from the jury'.

Traber retained ownership until it was consigned to auction at RM Sothebys' sale during the August 2015 Monterey Car Week. With an estimated selling price between $1.9m and $2.4m, the hammer descended at $2,062,500 including fees. Intriguingly, one of the 'Fageol' Fiat 8V Supersonics was auctioned by Bonhams during its Quail Lodge sale only a few days apart. That impeccably restored 1954 Geneva motor show car sold for $1,815,000 including premium.

The Jaguar Supersonic was sold to American-born Thai businessman, William E. 'Bill' Heinecke. The founder of Minor International, a global hospitality company and one of the largest restaurant and lifestyle companies in the Asia Pacific region. He has a long history of two- and four-wheel endeavour. An avid collector, racer and driver, the former Porsche

Jaguar XK120 Ghia Supersonic: the life and times of chassis 679768

• The original build plate; the donor silver/red XK 120 was dispatched to Delcroix, Paris on June 12 1952. It was delivered to marque concessionaire Malpelli and registered in July with the licence number 69 BJ 75 *Darin Schnabel.*

Carrera Cup Asia regular has spent more than half a century enjoying all things automotive. "I have been an auto enthusiast ever since I first began racing go-karts when I was 13 years-old," he said.

'I moved into competing in cars four years later aboard my first Mini Cooper S. When I was 18, I competed in the Macau Grand Prix with a Jaguar E-type. I moved on from that to an Elva-BMW and then a variety of other racing cars. I have collected cars and motorcycles since 1967 when I was still a teenager.
The Supersonic appealed to me because the first car I ever acquired was a Jaguar XK 120 drophead. As a result of that, the XK 120 has always been very special to me. The Ghia Supersonic is truly extraordinary. The styling is incredible, and I much prefer this version to the Fiat 8V-based Supersonics. When coupled with that Jaguar powertrain, the car is fantastic to drive. It has a great deal of power thanks to the Conrero-tuned engine.

Heinecke wasted little time relishing chassis 679768, and is eager to continue the car's illustrious record of participating in *concours d'elegance* events, one that stretches back to 1954. The Supersonic was displayed at the Windsor Castle Concours d'Elegance in September 2016 where it was one of 60 cars vying for honours. Competition spanned everything from an 1895 Benz Velo 'Comfortable' to the newly launched Touring Superleggera Alfa Romeo Disco Volante Spyder. Displayed alongside Jack and Debbie Thomas' unique 1955 Ferrari 375 America Coupé Speciale, it was one of the undoubted star turns in an event not exactly wanting for highlights.

That this dazzling machine is still wowing onlookers is remarkable given how many cars of this ilk have long since dropped off the radar. Either that, or been chopped, changed, silted and reconfigured. Chassis 679768 retains its original numbers-matching engine, complete with period tuning equipment, along with its Torinese coachwork. The interior remains as Ghia's artisans completed it back in 1954. Heinecke considers himself a caretaker as much an owner, and is eager to maintain – and use – the car as its creators intended. He said: "I hope to exhibit the Supersonic at some of the best shows and events around the world including the *concours* at Villa d'Este and Pebble Beach. Most of all, though, I want to enjoy using it."

This most characterful – and memorable – of Ghia creations is clearly in good hands.

Acknowledgements

About the author

Richard Heseltine is a journalist and author specialising in classic and contemporary performance cars and motor sport history. A former staff member of *Classic & Sports Car* and *Motor Sport* magazines, he has contributed extensively to other publications including *Octane*, *Classic Cars*, *Autosport*, *Motorsport News* and *Auto Italia*. He also writes for newspapers, notably *The Guardian*. Richard is the author of books on subjects as diverse as small-series British sports cars, Italian coachbuilding, Ferrari design history and a biography of racing driver and team principal Graham Warner. He lives in Shropshire where he attends to a coachbuilt Italian classic of his own, a 1966 Moretti 850 Sportiva.

Acknowledgements

Putting together a book – any book – is not the work of a moment. Nor does an author toil away in a vacuum. As such, I would like to thank my editor Giles Chapman, whose prose I have long admired, and whose knowledge of all things automotive is fathomless. I also need to thank the team at Porter Press, not least the boss himself, Philip Porter, and Tania Brown for their support. In addition, I would like to express my eternal gratitude to Margaret Heseltine, Bob Hui, Guy Allen and Rashed Chowdhury for riding to the rescue, and more than once. I would also like to express my thanks to the team at Motorsport Images (formerly LAT Photographic) who have been kindness itself, and also to Jaguar authority Terry McGrath. To those whose names I have omitted, you know who you are and you have my sincere appreciation.

- The owner of the Jaguar Supersonic, William E. Heinecke (centre), with Lennart P. Schouwenburg (left) and Jurriaan Schouwenburg of Strada e Corsa BV, the Netherlands-based restoration company responsible for the car's exquisite presentation today.

Index

Abarth, Carlo 43, 57, 68
Abarth cars, 64, 89
Agnelli, Gianni 31, 33, 51
Alberti, Giorgio 28
Alfa Romeo, 26, 28, 31, 43, 57, 70
Alfa Romeo cars 40
    6C 2500 27, 28, 29
    1900 7, 44, 52, 64
    Giulietta 43, 54
Alfa Romeo engines 52, 57, 66, 68
Allemano, Serafino 89
Allemano XK 140/XK **150** 89
(The) Autocar magazine 55, 66, 68, 73, 77

Beaumont, Count Mario Revellidi 26-27
Bertone/Carrozzeria Bertone 26, 27, 34, 36, 84
Bertone Jaguar XK **150** 84
Bianco, Fedele 28
Biondetti, Clemente 40, 78
Boano, Felice Mario 26-29, 44, 46, 58, 64-65
Boano, Gian Paolo 28, 64
Boano/Carrozzeria Boano 64, 88
Boano Lavorazioni Specialia 64
Boano (Loewy) XK 140 88-89
Boano SpA 64
Bonhams 86, 103, 104
Borrani wire wheels 13, 104
Bristol 87
Bugatti 26, 34

Cadillac 31, 45, 87
Cavagnero, Mario 68
Chrysler/Chrysler Corporation 26-27, 29, 33, 45, 49, 51, 58
Chrysler cars
    Crown Imperial limousine 33-34
    Dart show car 45, 47
    D'Elegance show car 46, 61
    Firebomb show car 33
    'Idea Cars' 29, 45
    Turbine Car 34, 47, 49-51
    K-3 show car 24, 29
Cisitalia/Consorzio Industriale Sportive Italia, 38 40-42, 44, 42, 87
Cisitalia cars
    202 40, 42-45, 70
    D46 40-41, 43
Classic & Sports Car magazine 77, 90, 93, 100, 101
Conrero, Virgilio 7, 43-45, 52, 54, 57, 66, 68
Conrero/Autotecnica Conrero 44, 52, 68, 90, 94, 96, 101, 105
Conrero cars
    Conrero Alfa Romeo 52, 54, 56, 66, 68, 77
Cowell, Richard C. 77

de Tomaso, Alessandro/Alejandro 30, 33-37, 54
De Tomaso Pantera 34, 36-37
Delahaye 28, 87
Detroit 7, 29, 34, 42, 45, 49, 50, 62, 76
Diatto 24
Dusio, Piero 38, 41, 43, 44, 52

Ellena/Carrozzeria Ellena 64
Ellena, Ezio 64
Engel, Elwood 49
Exner, Virgil 24, 29, 45-47, 49, 51, 58

Fageol, Lou 72, 74
Farina/Stabilimenti Farina 26, 64, 86, 95
    Stabilimenti Farina Jaguar XK **120** 86, 88
Farrago, Paul 58, 76
Fehlmann, Robert 52, 54, 66
Ferchaud, Jean-Claude 95, 100, 101, 103
Ferrari 31, 38, 47, 49, 54, 64, 66, 100, 105
Fiat 24, 26, 28, 31, 32, 34, 38, 40, 44, 51, 52, 57, 64, 66, 70, 73-73, 76
Fiat cars
    8V 7, 64, 68-74, 76-77, 103, 104, 105
    508 26-27
    600 30, 31, 33, 64
    1100 28, 33, 44, 58, 70, 89
    1500 26-28
    2300S 30, 33-34
Ford/Ford Motor Company 34, 36-37, 44, 58, 76, 89
Ford cars
    Fiesta 37
    Flashback show car 36
Ford, Henry II 43, 44,
Formula Junior 34, 54, 57
Formula One 54, 57, 94
Frua/Carrazzeria Frua 21, 33
Frua, Pietro 33, 45, 87

Gariglio, *Signor* 24
General Motors, 26, 74, 88
Geneva motor show, 70, 84, 88, 104
Ghia/Carrozzeria Ghia 7, 18, 34-47, 54-61, 58, 62, 64, 70, 75, 77, 90, 95, 101, 102, 105
Ghia, Giacinto 24, 26-27, 64
Ghia-Aigle/Carrosserie Ghia-Aigle 28, 87
Ghia-styled cars
    450SS 32, 34
    Alfa Romeo 6C 2500 '*fuoriserie*' 28–29
    C.230 prototype 34
    Chrysler Dart show car 24, 29, 45, 47
    Chrysler K-3 show car 24, 29
    Crown Imperial limousines 23–24
    De Tomaso Pantera 34, 36-37
    De Tomaso Vallelunga 36
    Fiat 8V 70, 77
    Gilda/Ghia-X 45-49
    Innocenti 950 Spider 33, 62-63
    Jaguars 86-87, 90, 92, 101
    Jolly variants 30-33
    L6.4 32, 33, 51
    Supersonic 7, 21-23, 24, 37, 45, 66-77, 102, 105
    Alfa Romeo 1900 Supersonic 52, 54, 56, 66, 68, 77
    Aston Martin DB2/4 Supersonic 7, 72, 73, 77
    Fiat 8V Supersonic 7, 66, 72, 74, 77, 103, 104, 105
    Shelby/AC Cobra 427 Supersonic 76
    Willment Supersonic hybrid 75–76

Jaguar XK 120 Supersonic 7, 77, 87, 90-105
Volkswagen Karmann-Ghia 31, 46-47, 58, 61, 62
Giacosa, Dante 38, 40, 69, 70
Giugiaro, Giorgetto 34, 36, 51

Hayworth, Rita 45, 47, 49
Heinecke, William E.'Bill' 8, 90-105
Hoffman, Max 88
Huebner, George 49, 51

Italian Design Studio (IDS) 37

Jager, Dirk de 103
Jaguar 8, 28, 77, 78, 80, 84, 85, 86, 90, 94
Jaguar cars
    C-type 94, 100, 102
    D-type 8, 83, 86, 95
    E-type 83, 84, 89, 100, 105
    MkVII 86, 95
    XK **120** 7, 8, 77, 78-83, 86, 87, 94, 103, 105
    Jaguar XK 120 Ghia Supersonic 23, 90-105
      chassis **679768** 8, 90-105
      Pinin Farina Jaguar XK 120 SE 88
      Stabilimenti Farina Jaguar XK 120 86
    XK **140** 82, 83, 85-89
    XK **150** 83, 84, 86, 87, 89, 100
Jaguar Drivers' Club 18, 94, 95, 103
John Willment Racing Organisation 76
Johnson, Leslie 78, 80-83
Johnson, Lyndon 33

Lambretta 30, 33
Lancia 26-27, 28, 31, 36, 54, 65, 70
Aprilia 27, 28
Aurelia B20 GT 64, 66, 68
Le Mans 24 Hours 8, 74, 82, 94, 95, 100
Loewy, Raymond 88-89
Louis Vuitton Concours d'Elegance 8, 101
Lyons, William 78, 80, 83, 86, 90

Mairesse, Guy 95
Malpelli, *Monsieur* 90, 92, 93, 105
Martinengo, Franco 86
Martínez, Rafael Leónidas Trujillo 34
Maserati 34, 70, 87, 89
Mercury Monarch 37, 89
Michelotti, Giovanni 28, 45, 54, 57, 86-87, 89
Michelotti Jaguar XK 140 SE 86-87
Mille Miglia 7, 40, 52, 54, 58, 66, 68, 72, 80
Monviso/Stablimento Monviso 31
Monza 38, 46
Moro, Giacomo Gaspardo
Morocco, King of 29, 45
Moss, Stirling 78, 80, 81, 83
Moto Guzzi engine 43, 46
Motor Italia magazine 26, 28
Motto, Rocco 87
Motto Jaguar XK 150 87-88

New York Auto Show 36, 49, 74
Nuvolari, Tazio 40

Officine Elettromeccaniche 43, 52
Opel 57
OSCA engine 46, 49
OSCA MT4 100
OSI (Officine Stampaggi Industriali) 33-34, 62

Paris motor show 74, 84, 89
Pebble Beach concours 102-103, 104, 105,
Pinin Farina 26-29, 42, 62, 64, 73, 86, 88
Pinin Farina Jaguar XK 120 SE 88
Pininfarina 36
Plymouth cars 27, 29, 32, 34, 58
    XX-500 prototype 27, 29, 58
Pollo, Luciano 64
Politecnico di Torino 44, 46, 51
Porsche 28, 76
Porsche, Ferdinand 43

RAC Rally of Great Britain 78
Rapi, Fabio Lucio 69-70, 73
Renault 30, 33, 54, 57, 87
Renault, Dr Philippe 93, 94
Rétromobile, 95, 101
*Rétroviseur m*agazine 95, 100
Road & Track magazine 66, 74, 88
Royal-Elysées 77, 90, 92, 93

Sapino, Filippo 37, 58
Savonuzzi, Giovanni 7, 11, 31, 33, 38-51, 52, 68, 76, 87, 94
Scaglione, Franco 84
Schell, Harry 73, 77
Segre, Luigi 'Gigi' 28-34, 44-46, 51, 58-62, 64, 68
Shelby Cobra 31, 76
Siata 58, 70, 73
Simca 26, 54
Sothebys 77, 104
SVA (Società Valdostana Automobili) 43, 52, 87

Targa Florio 24, 57
Tjaarda, Tom 33-37, 58, 63
Traber, Christian 103, 104
Trujillo, Rafael Leónidas 34, 36
Turin motor show 31, 46, 49, 68, 73, 77, 84

Urban, Roland 86, 94, 95, 100

Vack, Peter 49, 51
Vignale/Carrozzeria Vignale 37, 44, 66, 69, 70, 73, 76
Vignale, Alfredo 37, 42
Viotti/Carrozzeria Viotti 26-27, 64
Volkswagen 62

Walker, Peter 78, 100
Weber carburettors 70, 90, 92, 102
Wilke, Bob 47, 49

Zagato 68, 73, 76
Zagato Jaguar XK 140/XK 150 84-86